Advances in
Electronic Ceramics

Advances in Electronic Ceramics

A Collection of Papers Presented at the 31st International Conference on Advanced Ceramics and Composites
January 21–26, 2007
Daytona Beach, Florida

Editors
Clive Randall
Hua-Tay Lin
Kunihito Koumoto
Paul Clem

Volume Editors
Jonathan Salem
Dongming Zhu

The American Ceramic Society

BICENTENNIAL
1807
WILEY
2007
BICENTENNIAL

WILEY-INTERSCIENCE
A John Wiley & Sons, Inc., Publication

Published by John Wiley & Sons, Inc., Hoboken, New Jersey.
Published simultaneously in Canada.

For general information on our other products and services or for technical support, please contact our
Customer Care Department within the United States at (800) 762-2974, outside the United States at
(317) 572-3993 or fax (317) 572-4002.

Wiley also publishes its books in a variety of electronic formats. Some content that appears in print may
not be available in electronic format. For information about Wiley products, visit our web site at
www.wiley.com.

Wiley Bicentennial Logo: Richard J. Pacifico

Library of Congress Cataloging-in-Publication Data is available.

ISBN 978-0-470-19639-7

Printed in the United States of America.

10 9 8 7 6 5 4 3 2 1

Contents

ELECTROCERAMIC MATERIALS FOR SENSORS

THERMOELECTRIC MATERIALS FOR POWER
CONVERSION APPLICATIONS

TRANSPARENT ELECTRONIC CERAMICS

Preface

This issue contains a collection of 27 papers on electronic ceramic materials and technology that were presented during the 31st International Conference on Advanced Ceramics and Composites, Daytona Beach, FL, January 21–26, 2007. Papers were submitted from the following Symposia and Focused Sessions:

- Symposium on Advanced Dielectric, Piezoelectric and Ferroelectric Materials
- Electroceramic Materials for Sensors
- International Symposium on Thermoelectric Materials for Power Conversion Applications
- Transparent Electronic Ceramics

The editors wish to extend their sincere gratitude and appreciation to all the authors for their cooperation and contribution to this proceedings as well as to the reviewers. Thanks also are due to all of the participants, organizers, and session chairs. The attendees were highly international and there were very good technical exchanges and interactions between the participants. We look forward to future symposia of similar standards at a forthcoming "Daytona Beach" meeting.

We hope that this issue will serve as a useful reference for researchers and technologists working in the field of advanced electronic ceramics.

CLIVE RANDALL
Pennsylvania State University

HUA-TAY LIN
Oak Ridge National Laboratory

KUNIHITO KOUMOTO
Nagoya University

PAUL CLEM
Sandia National Laboratory

Introduction

2007 represented another year of growth for the International Conference on Advanced Ceramics and Composites, held in Daytona Beach, Florida on January 21-26, 2007 and organized by the Engineering Ceramics Division (ECD) in conjunction with the Electronics Division (ED) of The American Ceramic Society (ACerS). This continued growth clearly demonstrates the meetings leadership role as a forum for dissemination and collaboration regarding ceramic materials. 2007 was also the first year that the meeting venue changed from Cocoa Beach, where it was originally held in 1977, to Daytona Beach so that more attendees and exhibitors could be accommodated. Although the thought of changing the venue created considerable angst for many regular attendees, the change was a great success with 1252 attendees from 42 countries. The leadership role in the venue change was played by Edgar Lara-Curzio and the ECD's Executive Committee, and the membership is indebted for their effort in establishing an excellent venue.

The 31st International Conference on Advanced Ceramics and Composites meeting hosted 740 presentations on topics ranging from ceramic nanomaterials to structural reliability of ceramic components, demonstrating the linkage between materials science developments at the atomic level and macro level structural applications. The conference was organized into the following symposia and focused sessions:

- Processing, Properties and Performance of Engineering Ceramics and Composites
- Advanced Ceramic Coatings for Structural, Environmental and Functional Applications
- Solid Oxide Fuel Cells (SOFC): Materials, Science and Technology
- Ceramic Armor
- Bioceramics and Biocomposites
- Thermoelectric Materials for Power Conversion Applications
- Nanostructured Materials and Nanotechnology: Development and Applications
- Advanced Processing and Manufacturing Technologies for Structural and Multifunctional Materials and Systems (APMT)

- Porous Ceramics: Novel Developments and Applications
- Advanced Dielectric, Piezoelectric and Ferroelectric Materials
- Transparent Electronic Ceramics
- Electroceramic Materials for Sensors
- Geopolymers

The papers that were submitted and accepted from the meeting after a peer review process were organized into 8 issues of the 2007 Ceramic Engineering & Science Proceedings (CESP); Volume 28, Issues 2-9, 2007 as outlined below:

- Mechanical Properties and Performance of Engineering Ceramics and Composites III, CESP Volume 28, Issue 2
- Advanced Ceramic Coatings and Interfaces II, CESP, Volume 28, Issue 3
- Advances in Solid Oxide Fuel Cells III, CESP, Volume 28, Issue 4
- Advances in Ceramic Armor III, CESP, Volume 28, Issue 5
- Nanostructured Materials and Nanotechnology, CESP, Volume 28, Issue 6
- Advanced Processing and Manufacturing Technologies for Structural and Multifunctional Materials, CESP, Volume 28, Issue 7
- Advances in Electronic Ceramics, CESP, Volume 28, Issue 8
- Developments in Porous, Biological and Geopolymer Ceramics, CESP, Volume 28, Issue 9

The organization of the Daytona Beach meeting and the publication of these proceedings were possible thanks to the professional staff of The American Ceramic Society and the tireless dedication of many Engineering Ceramics Division and Electronics Division members. We would especially like to express our sincere thanks to the symposia organizers, session chairs, presenters and conference attendees, for their efforts and enthusiastic participation in the vibrant and cutting-edge conference.

ACerS and the ECD invite you to attend the 32nd International Conference on Advanced Ceramics and Composites (http://www.ceramics.org/meetings/daytona2008) January 27 - February 1, 2008 in Daytona Beach, Florida.

JONATHAN SALEM AND DONGMING ZHU, Volume Editors
NASA Glenn Research Center
Cleveland, Ohio

Advanced Dielectric, Piezoelectric and Ferroelectric Materials

LEAD STRONTIUM ZIRCONATE TITANATE (PSZT) THIN FILMS FOR TUNABLE DIELECTRIC APPLICATIONS

Mark D. Losego and Jon-Paul Maria
The Electroceramic Thin Film Group, North Carolina State University
1001 Capability Drive
Raleigh, NC 27606

ABSTRACT
    Lead strontium zirconate titanate (PSZT) solid solutions are investigated for low-process temperature, tunable dielectric applications. PSZT films of three compositions (40%, 50% and 55% Sr) are prepared via a chemical solution deposition route. All films display paraelectric properties at room temperature and dielectric losses below 0.05 under field. Dielectric tunability is demonstrated to increase with annealing temperature. However, even films crystallized at 650°C have tunabilities of at least 24% at 400 kV/cm, with films of the $(Pb_{0.45}Sr_{0.55})(Ti_{0.7}Zr_{0.3})O_3$ composition exhibiting a tunability of 57%. Such dielectric tunability approaches typical values for low-temperature deposited $(Ba,Sr)TiO_3$ films crystallized at 900°C.

INTRODUCTION
    The dielectric constant of ferroelectric materials can be varied by the application of an external electric field. Thin films of such tunable dielectric materials are of interest in microelectronic devices used for communication. Recent examples include tunable bandpass filters and phase shifters based on barium strontium titanate (BST) thin films.[1,2] For such applications, it is desirable to choose material compositions with ferroelectric transitions below room temperature such that the device operates in the paraelectric state. Operating tunable dielectrics in the paraelectric regime offers two basic advantages. First, device design is simplified because dielectric tuning is non-hysteretic. Second, dielectric losses are reduced because domain walls are absent.
    Currently, most thin film ferroelectric varactors employ solid solution compositions of the $(Ba,Sr)TiO_3$ perovskite family. However, because of its refractory nature, this materials set requires high processing temperature (>800°C) to achieve sufficient grain size and crystallinity for desirable dielectric tunabilities. Because certain substrate materials of interest for frequency agile devices, like high-resistivity silicon, require lower thermal budgets, tunable dielectrics that can be processed at lower temperatures warrant investigation. Recently, Lu and coworkers[3,4] have examined a bismuth-based pyrochlore that can be processed at 750°C. However, because this material is not ferroelectric, it requires a large electric field (2400 kV/cm) to achieve reasonable dielectric tuning (55%). We propose to investigate one of the most common low-process temperature ferroelectrics, $Pb(Zr,Ti)O_3$, as a possible candidate for tunable dielectric applications. Specifically, we choose to modify the A-site of PZT with Sr to achieve a room temperature paraelectric state material and examine the effect this has on the thin film's processability, dielectric response, and tunability.

3

EXPERIMENTAL PROCEDURE

Compositional selection of $(Pb,Sr)(Ti,Zr)O_3$ films was based upon the room temperature quaternary bulk phase diagram,[5] which is shown in Fig. 1. Compositions with near room temperature phase transitions were chosen in an attempt to retain high tunabilities. Three different PSZT compositions were investigated: $(Pb_{0.5}Sr_{0.5})(Ti_{0.5}Zr_{0.5})O_3$, $(Pb_{0.6}Sr_{0.4})$ $(Ti_{0.4}Zr_{0.6})O_3$, and $(Pb_{0.45}Sr_{0.55})(Ti_{0.7}Zr_{0.3})O_3$. These compositions are labeled in the phase diagram of Fig. 1. A $Pb(Zr_{0.52}Ti_{0.48})O_3$ film was also prepared as a standard for comparison.

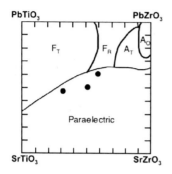

Fig. 1: Room temperature quaternary phase diagram for the $(Pb,Sr)(Ti,Zr)O_3$ system.[5] Black dots represent the three compositions investigated in this paper. Phases fields are notated as F for ferroelectric and A for antiferroelectric; crystal structure is notated by subscripts: T for tetragonal, O for orthorhombic, and R for rhombohedral.

Films were prepared using a chemical solution deposition approach. To prepare the solution, titanium isopropoxide and zirconium propoxide (70 wt% in 1-propanol) precursors were reacted in proper stoichiometric ratio in a low-humidity (<20%) environment. Acetic acid was added in a 4:1 molar ratio with the transition metal cations. After 10 minutes of continuous stirring, a 3:1 molar ratio of acetylacetone was added to chelate the metal cations. At this point the solution could be removed from the dry environment. The solution was diluted to ~0.4 M with acetic acid. Next, strontium acetate was added. The solution was heated to 90°C to facilitate dissolution; the strontium precursor took ~35 min to completely dissolve. Finally lead acetate tri-hydrate was added to give a 10 mol% excess A-site stoichiometry to account for probable lead loss during film crystallization. Because Pb was the only volatile constituent, film stoichiometry was assumed to reflect the Sr:Ti:Zr ratio in the prepared solution. The solution was then further diluted with acetic acid to achieve a 0.3 M concentration. Solutions exhibited limited stability with visible precipitation within 2-3 days. Solution stability could be extended to approximately one week if the solution was refrigerated at 0°C.

Films were prepared by spin-coating the solutions on platinized silicon substrates at 3000 rpm for 30 seconds. Films were dried on a hotplate at 325°C for 5 min. This process was repeated for three layers. The PSZT films were then crystallized in a tube furnace at a range of temperatures (600°C-800°C) for 30 min. Final film thickness was ~320 nm.

The crystalline structure of the PSZT films was verified by x-ray diffraction using a Bruker AXS D-5000 with an area detector. Atomic force microscopy was employed to image surface morphology; images were collected in tapping mode. Dielectric properties were probed with an HP4192a impedance analyzer. Besides examining the dielectric tunability, measurements were made as a function of temperature to confirm the Curie point shift. All dielectric measurements were made at 10 kHz with a 0.05 V oscillating voltage. Dielectric tunability is calculated as $(\kappa_{max} - \kappa_{min})/(\kappa_{max})*100\%$ where κ is the dielectric constant.

RESULTS AND DISCUSSION

X-ray diffraction data for three different PSZT compositions fired at 650°C is presented in Fig. 2. These samples have a phase pure perovskite structure with no indication of pyrochlore crystallites. Fig. 3 provides a sampling of the surface morphology for these films; an image of a PZT film prepared at the same temperature (650°C) is included for comparison. Whereas the PZT film exhibits a uniform morphology indicative of columnar grains nucleating at the substrate, the PSZT films contain surface nucleated grains forming rosette structures.[6] Currently, it is unclear whether variations in solution chemistry and/or film processing conditions could improve the uniformity of this microstructure or whether the observed surface nucleation is a direct result of strontium lowering the activation energy for nucleation in this system.

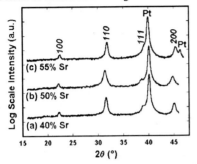

Fig. 2: X-ray diffraction data for chemical solution deposited PSZT films of varying composition annealed at 650°C. Perovskite peaks are labeled by reflection while Pt peaks are observed from the substrate.
(a) $(Pb_{0.6}Sr_{0.4})(Ti_{0.4}Zr_{0.6})O_3$
(b) $(Pb_{0.5}Sr_{0.5})(Ti_{0.5}Zr_{0.5})O_3$
(c) $(Pb_{0.45}Sr_{0.55})(Ti_{0.7}Zr_{0.3})O_3$

Fig. 3: Atomic force microscopy images of chemical solution deposited PSZT films crystallized at 650°C. Images are 2 μm x 2 μm:

(a) $Pb(Zr_{0.52}Ti_{0.48})O_3$
(b) $(Pb_{0.6}Sr_{0.4})(Ti_{0.4}Zr_{0.6})O_3$
(c) $(Pb_{0.5}Sr_{0.5})(Ti_{0.5}Zr_{0.5})O_3$
(d) $(Pb_{0.45}Sr_{0.55})(Ti_{0.7}Zr_{0.3})O_3$

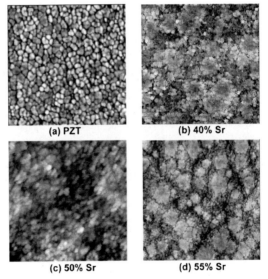

(a) PZT

(b) 40% Sr

(c) 50% Sr

(d) 55% Sr

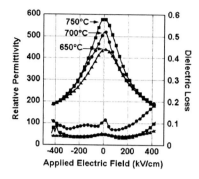 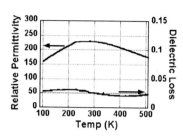

Fig. 4: Room temperature dielectric tunability at 10 kHz sampling frequency for three $(Pb_{0.45}Sr_{0.55})(Ti_{0.7}Zr_{0.3})O_3$ thin films crystallized at varying temperatures.

Fig. 5: Measurement of dielectric properties as a function of temperature for a $(Pb_{0.5}Sr_{0.5})(Ti_{0.5}Zr_{0.5})O_3$ thin film collected at 10 kHz.

Various annealing temperatures were investigated to determine the effect on dielectric tunability. As expected, tunability increased with increasing annealing temperature—a result of improved crystallinity and larger grain size. An example of this behavior is presented in Fig. 4 for the 55% Sr PSZT film fired at 650°C, 700°C, and 750°C. For this film, dielectric tunability increases from 57% at 650°C to 67% at 750°C.

The shape of the curves presented in Fig. 4 indicates material in a paraelectric state at room temperature—dielectric tuning with negligible hysteresis and peak permittivity at zero bias. To further confirm that the Curie point has been shifted to below room temperature, dielectric measurements were collected as a function of temperature. An example of temperature dependent dielectric data is presented in Fig. 5 for a PSZT film of the 50% Sr composition. The phase transition for this film appears to occur over a broad temperature range, common for most ferroelectric thin films. However the transition regime is clearly centered at or slightly below room temperature (~298 K)—approximately 350°C lower than unmodified PZT of the same composition. Thus, we verify that the addition of strontium has created a material in the paraelectric state at room temperature.

Table I: Summary of dielectric properties for PZT and PSZT films crystallized at 650°C and measured at room temperature and 10 kHz. Dielectric tunability is taken at 400 kV/cm

Composition	Maximum κ	Average tanδ	Dielectric Tunability
$Pb(Zr_{0.52}Ti_{0.48})O_3$	1450	0.022	80%
$(Pb_{0.6}Sr_{0.4})(Ti_{0.4}Zr_{0.6})O_3$	220	0.023	24%
$(Pb_{0.5}Sr_{0.5})(Ti_{0.5}Zr_{0.5})O_3$	220	0.033	39%
$(Pb_{0.45}Sr_{0.55})(Ti_{0.7}Zr_{0.3})O_3$	440	0.047	57%

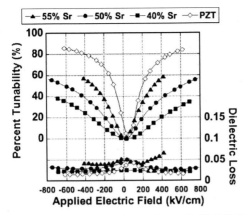

Fig. 6: Plot of dielectric tunability and dielectric loss under applied DC field for PSZT and PZT thin films crystallized at 650°C. Measurements taken at 10 kHz and room temperature.

Finally, in Fig. 6 dielectric tunability data is presented for all three PSZT compositions with pure PZT as a reference. Table I summarizes the dielectric properties of these films. All films presented in this plot are crystallized at 650°C—a relatively low temperature compared to typical BST thin film varactors (800°C – 900°C).[7] Of the investigated compositions, the 55% Sr system appears the most tunable with 57% tunability at 400 kV/cm. Typical polycrystalline BST films prepared at ~900°C have tunabilities of ~65% at 400 kV/cm.[7] Values for dielectric loss also appear promising in this system, largely remaining below 0.05. Further improvements in dielectric loss could be expected from Pb content optimization and acceptor doping.

CONCLUSIONS

Phase pure PSZT films have been prepared via a chemical solution route on platinized silicon substrates. The structure and dielectric response of these films is compared to a similarly prepared PZT film. In contrast to the uniform microstructure of the PZT film, PSZT films have an inhomogeneous structure with features resembling surface-nucleated rosettes. Despite this microstructure, these films exhibit relatively low dielectric losses (<0.05) and reasonable dielectric constants (ranging from 210 to 440 when crystallized at 650°C). Of the compositions investigated, the $(Pb_{0.45}Sr_{0.55})(Ti_{0.7}Zr_{0.3})O_3$ film exhibits the highest dielectric tunability (57% at 400kV/cm) when crystallized at 650°C. This tunability approaches typical values reported for BST thin films prepared at 900°C (~65% at 400kV/cm) but is significantly lower than the pure PZT film (80% at 400kV/cm).

REFERENCES
[1]J. Nath *et al.* An Electronically Tunable Microstrip Bandpass Filter Using Thin-Film Barium Strontium Titanate (BST) Varactors, *IEEE Transactions on Microwave Theory and Techniques*, **53**, 2707 (2005).
[2]J. L. Serraiocco, P. J. Hansen, T. R. Taylor, J. S. Speck, and R. A. York, Compact Distributed Phase Shifters at X-band Using BST, *Int. Ferroelec.*, **56**, 1087-1095 (2003).

[3]J. Lu, S. Stemmer, Low-loss, Tunable Bismuth Zinc Niobate Films Deposited by RF Magnetron Sputtering, *Appl. Phys. Let.*, **83**, 2411-3 (2003).

[4]J. Lu, *et al.* Low-loss tunable capacitors fabricated directly on gold bottom electrodes, *Appl. Phys. Let.,* **88**, 112905 (2006).

[5]T. Ikeda, A Few Quarternary Systems of Perovskite Type $A^{2+}B^{4+}O_3$ Solid Solutions, *J. Phys. Soc. Jap.*, **14,** 1286-94 (1959).

[6]B. A. Tuttle *et al.* Microstructural Evolution of $Pb(Zr,Ti)O_3$ Thin-Films Prepared by Hybrid Metalloorganic Decomposition, *J. Mater. Res.,* **7,** 1876-1882 (1992).

[7]D. Ghosh *et al.* Tunable High-Quality-Factor Interdigitated $(Ba,Sr)TiO_3$ Capacitors Fabricated on Low-cost Substrates with Copper Metallization, *Thin Solid Films,* **496,** 669-673 (2006).

NANOSIZE ENGINEERED FERROELECTRIC/DIELECTRIC SINGLE AND MULTILAYER FILMS FOR MICROWAVE APPLICATIONS

R. Wördenweber, E. Hollmann, Mahmood Ali, J. Schubert, and G. Pickartz
Institut für Schichten und Grenzflächen (ISG) und cni - Center of Nanoelectronic Systems
for Information Technology
Forschungszentrum Jülich , D-52425 Jülich, Germany

ABSTRACT

The possibility of using ferroelectric based layered structures is very attractive for both basic research and application. Combining ferro- and dielectric layers could lead to improvements of the dielectric properties, it might lead to a possible engineering of the ferroelectric properties. In this paper we report on the effect of strain induced modification of the ferroelectric properties of $SrTiO_3$ single layers and $SrTiO_3/CeO_2$ multilayers. The films are deposited via pulsed laser deposition technology on microwave suitable substrates (i.e., r-cut sapphire) with thicknesses ranging from 8nm to 710nm. The crystallographic structures and electronic properties of the single and multilayers are analysed and compared with the properties of the unstrained single crystalline $SrTiO_3$. Due to the large lattice mismatch and the difference in thermal expansion between the carrier and the film, large compressive strain is induced in the ferroelectric film that leads to induced ferroelectricity in the $SrTiO_3$ layers that persists up to 200-220K. This enhancement is accompanied by a considerable enhancement of the permittivity and tunability even at room temperature. Considering the advantage of the use of sapphire as a low-loss microwave material, this makes this system an interesting system for tunable microwave devices like varactors or phase shifters.

INTRODUCTION

In the past, $SrTiO_3$ (STO) has been of considerable interest due to its low temperature properties. Nowadays the interest has revived due to new insight that allows to modify or even engineer the properties towards various electronic room-temperature applications in ferroelectronic devices (e.g., spin-based electronics) or tunable devices (e.g., varactors or phase shifters for microwave applications).

Generally, the dielectric properties of STO are qualitatively similar to those of the paraelectric phase of typical perovskite ferroelectrics like $BaTiO_3$. In these materials exist a soft transverse optical mode whose frequency tends to zero with decreasing temperature [1,2]. This leads to an increase of the permittivity of STO when the material is cooled [3]. The temperature dependence of the permittivity (and soft-mode frequency) follows the Curie-Weiss law with a Curie temperature of about 40K (see also fig. 2) [4]. In contrast to $BaTiO_3$, STO does not undergo a ferroelectric transition. This is attributed to quantum fluctuations in the material at low temperatures that suppress the ferroelectric transition. Therefore, STO is called an incipient ferroelectric or a quantum paraelectric [3,5,6]. This quantum paraelectric state is very sensitive to perturbations of the lattice. A ferroelectric transition can been induced in STO for instance by small levels of impurities or doping [7], applied electric fields [8], ^{18}O substitution [9], or mechanical stress [10,11]. As a result the Curie temperature can be increased and the material is now suitable for room-temperature applications. This is one of the major reasons for the renewed interest in this material.

One of the most direct methods to introduce ferroelectricity into STO is doping of STO with other cations. This has been demonstrated most clearly on doping STO with Ca, Bi, or Ba [7]. However, in case of thin film applications mechanical strain inducted to STO films by the carrier (substrate) offers a natural method to engineer the ferroelectric properties of

STO. Though hydrostatic pressure does not affect the quantum paraelectric state [12], uniaxial pressure was found to induce a ferroelectric phase [10]. This is attributed to the electrostrictive effect producing an electric field that induces a ferroelectric transition [10]. More recently the effects of biaxial strain in epitaxial thin films were analyzed theoretically by Pertsev *et al.* [13]. They predicted from Landau-Ginsburg-Devonshire theory that due to the coupling between the strain and the polarization in STO it was possible to induce ferroelectricity in epitaxial STO films. These predictions showed that true ferroelectric phases are present for films under sufficient compressive and tensile strains, respectively. By minimization of the Helmholtz free energy it was even possible to predict the temperature of phase transition as function of bi-axial strain. These predictions was confirmed for the tensile case of epitaxial STO films grown on $DyScO_3$ substrates [14] and, likewise, for compressively strained STO grown on $LaAlO_3$ [15] and CeO_2/Al_2O_3 [16]. However, a number of question remained, like

- is the strain purely defined by the lattice mismatch between STO and the substrate lattice,
- what is effect of the expansion coefficient, that is usually different for the film and the substrate, and
- is it possible to engineer the ferroelectric properties for instance by combining STO layers with other layers (e.g., dielectric or ferroelectric layers) of different lattice constant and thermal expansion coefficient.

In this paper we discuss some of these questions. Furthermore, we analyze the effect of the layer thickness on the structural properties (e.g., relaxation of the strain) and the resulting ferroelectric properties. Furthermore, we use microwave suitable system, i.e., CeO_2 buffered sapphire, as carrier for the STO films.

EXPERIMENTAL RESULTS AND DISCUSSION

A series of STO single layer and multilayer films with thickness of the STO layers ranging from 8nm to 710nm is grown on CeO_2 buffered r-cut sapphire (Al_2O_3). In order to provide identical growth conditions for the STO layers, all CeO_2 buffer layers are deposited at identical conditions (i.e., magnetron sputter technology with rf-power of 130 W on a 6" cathode, Ar/O_2 gas mixture of ratio 6.6/1 at a pressure of 13 Pa, and a heater temperature of 850°C resulting in a deposition rate of 0.5 nm/min). The thickness of the CeO_2 buffer layer is 40nm. Subsequently the STO layers are deposited via pulsed laser deposition at identical conditions (i.e., in pure O_2 atmosphere at a pressure of 1 Pa, laser power of about 5 J/cm^2 at the target, and 10 Hz repetition rate, resulting in a growth rate of 180nm/min).

XRD analysis:

The crystalline structure of the films is analyzed via X-ray diffraction analysis of the epitaxial growth and quality, the lattice parameters, and the mutual angular orientation of the different layers.

First, all STO films grow epitaxially. No STO phases other than (100) orientation are detected. The full width half maximum (FWHM) of the (200) reflex measured in Bragg-Brentano and rocking geometry are shown in fig. 1. The values are almost constant for larger thicknesses h_f of the STO layer, they increase for $h_f < 100$nm. The increase of the FWHM is characteristic for XRD measurements on thin samples. Together with the fact that the FWHM data of Bragg-Brentano and rocking curve behave very similar, this indicates that the STO film quality is comparable for all film thicknesses.

Second, the lattice parameter is obtained from the XRD measurements in Bragg-Brentano geometry using the Nelson-Riley correction of (l00) peaks. The STO lattice parameter normal to the film surface, $a_\perp$, shows no thickness dependence (see fig. 1). It is only slightly larger than the literature value of $a_{STO} = (3.905\pm0.002)$Å for unstrained single crystalline cubic STO. Assuming a conservation of the volume of the STO unit cell, the average lattice parameter of our films of $a\perp = (3.911 \pm 0.0015)$Å results in an extremely small tetragonality of (0.234 ± 0.057) %. Direct determination of the in-plane lattice parameter via XRD measurements of the (211) and (111) reflexes confirms this result. However, the accuracy of these direct measurements is not good enough to reproduce the exact value.

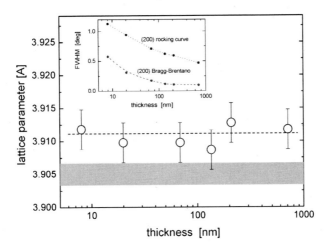

Fig. 1: Thickness dependence of the lattice parameter $a\perp$ normal to the substrate surface and of the FWHM of the rocking curve and Bragg-Brentano diffraction peaks (inset). The literature value ($a_{STO} = (3.905\pm0.002)$ Å) is indicated by the shaded area, the average lattice parameter is given by the dashed line.

Finally, the mutual structural orientations of the epitaxial grown layers and the substrate are examined. The largest lattice mismatch exists between the substrate (r-cut sapphire) and the CeO_2 buffer layer. As a result of the large lattice mismatch the layers grow under a tilt angle of 1-3° with respect to the orientation of the sapphire. A tilting of the STO orientation with respect to the CeO_2 buffer layer is not measured within the experimental resolution limit of about 0.05°.

From these measurements we conclude:

(i)   STO and CeO_2 grow epitaxially and without any tilting on top of each other.

(ii)   The lattice mismatch between the substrate and the film is mainly compensated by a tilting of the crystallographic orientation of the films with respect to the substrate.

(iii)   The experimentally determined tetragonal distortion of the STO is extremely small (i.e., much smaller than the theoretically expected tetragonality of about 2% for STO on CeO_2). This could be explained by a small tilting between the epitaxial layers which could not be resolved. More likely it has to be ascribed to other effects like dislocations, defects, oxygen deficiencies in the STO layer.

However, it should be noted, that all XRD measurements are performed at room temperature. The films are deposited and formed (STO phase formation) at much higher temperatures. Moreover, the electric properties are determined at low temperatures (4-300K). Therefore, additional effects of the different thermal expansion coefficients of the different components (substrate, buffer, and STO) have to be taken into account. This will considered for in the discussion of the electric measurements below.

Characterization of the ferroelectric properties:

The dielectric properties of the films are determined by capacitance measurements of planar capacitors at 1MHz and 2-4GHz. Furthermore, a comparison of the low frequency data and measurements up to the THz regime [17] demonstrates, that the permittivity is independent of frequency up to about 0.5 THz for our samples. Therefore, the dielectric properties are analyzed via capacitance measurements at 1MHz using a parallel capacitor model [18]:

$$C_f = C_{tot} - \left(C_s + C_{air} + C_{setup} \right), \tag{1}$$

that models the measured total capacitance C_{tot} by a parallel arrangement the contributions from the STO film (C_f), the substrate (C_s), the air above the planar capacitor (C_{air}), and the experimental setup (C_{setup}). The expression for the permittivity of the STO film is [16]:

$$\varepsilon_f = \frac{2C_f}{\varepsilon_0 \cdot F_f \cdot w} + \varepsilon_s \cong \frac{2C_f}{\varepsilon_0 \cdot F' \cdot h_f \cdot w} \tag{2}$$

with the vacuum permittivity ε_o, the dielectric constant of the substrate ε_s, the thickness h_f of the STO single layer or multilayer, the length of the electrodes w, and the design dependent constant F_f which can be approximated by $F_f \cong F'd_f$ for thin films [16]. $F'=0.4765\mu m^{-1}$ for our configuration (capacitor gap of 4μm and width of the electrodes of 500μm).

Fig.2 shows a comparison of the temperature and thickness dependence of the permittivity (at zero bias voltage) for the different STO single layers of different thickness and, for comparison, STO single crystal.

First, the room-temperature permittivity measured for STO films with a thickness $h_f >$ 100nm is similar to that of the single crystal, i.e. $\varepsilon(300K) \approx 350$. However, for STO films with $h_f < 100$nm the permittivity increases with decreasing thickness. For our thinnest sample ($h_f = 8$nm) we measured $\varepsilon(300K) \approx 790$ which is more than twice as large compared to the value obtained for single crystals.

Second, the STO single crystal and the thicker STO films ($h_f > 100$nm) show perfect Curie-Weiss behavior for higher temperatures:

$$\varepsilon = \frac{C_{CW}}{T - T_c}, \tag{3}$$

with a Curie-Weiss temperature $T_c \approx 40$K comparable to the literature values for bulk STO. Moreover, even a quantitative agreement of the permittivity ε of single crystals and thick films is given for high temperatures. Only for T < 150K the data diverge. With decreasing temperature the permittivity of the single crystal outranges the permittivity of the thick films by far.

Finally, with decreasing thickness of the STO layers ($h_f < 100$nm), the high temperature behavior of the permittivity diverges more and more from the Curie-Weiss behavior of the single crystalline STO. Whereas the film with a thickness of $h_f = 60$nm still shows Curie-Weiss behavior with a T_c comparable to that of bulk material (but with enhanced permittivity), the temperature dependence of ε^{-1} of the thinnest films ($h_f < 25$nm) are not Curie-Weiss like. Furthermore, the curves obtained for the thinner samples show a clear maximum around room temperature. The position of this maximum shifts to lower temperatures with decreasing film thickness.

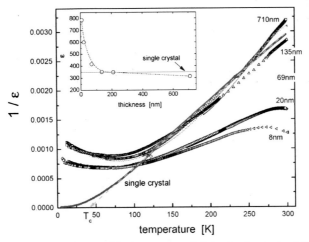

Fig. 2: Temperature dependence of the inverse permittivity for the strained STO films of different thickness and an unstrained STO single crystal for zero electrical bias field.

The difference in (a) the absolute value of $\varepsilon(300K)$ (see inset of fig. 2) and (b) the temperature dependence of ε^{-1} obtained for different film thicknesses can be explained by the following scenario:

- The strain is relaxed over the STO film thickness. This explains the quantitative and qualitative (Curie-Weiss behavior) agreement of the permittivities of thick films ($h_f >$ 100nm) and single crystals at elevated temperatures (T >150K).
- Only for thinner films ($h_f <$ 100nm) the strain can not relax which results in an increase of the permittivity at room temperature.
- The deviation from the Curie-Weiss behavior and, especially, the maximum in the temperature dependence of ε^{-1} observed for thin films indicates that the strain is not only induced by the lattice mismatch between the carrier (CeO_2/Al_2O_3) and the STO layer. It is more likely that the difference in thermal expansion between the substrate and the STO film leads to a temperature dependent strain in the film. This is consistent with the XRD observation (i.e., the small distortion of the STO). It can also explain the deviation of the permittivity observed for thick films at low temperatures.

Ferroelectric phase transition and tunability:

In contrast to the single crystal, all thin films show a large tunability and ferroelectricity up to elevated temperatures. The hysteretic behavior of the current-voltage characteristics is a clear fingerprint of the presence of ferroelectricity. Thus, the presence of ferroelectricity can for instance be analyzed by measurements of the slope dC/dV at zero voltage obtained after applying a large positive or negative bias fields, respectively. For hysteric behavior (i.e., ferroelectricity) a non-zero value of dC/dV should be measured.

Fig. 3 shows the temperature dependence of the normalized slope dC/dV for our films and, for comparison, an unstrained STO single crystal. In contrast to the single crystal, all films show clear indications for the presence of ferroelectricity up to 200-220K. The temperature dependence of the slope dC/dT is very similar for all samples. Therefore, we conclude, that the phase transition from the paraelectric to the ferroelectric phase takes place at T_p = 200 – 220 K for our STO films. The value obtained from this analysis for T_p agrees perfectly with the theoretical predictions [13]. However, it should be noted, that the theoretical predictions taken into account only the lattice mismatch between the carrier (here the buffer layer) and STO. The strain that is induced by the difference in thermal expansion between the substrate and the STO film is not considered in this model. Furthermore, it should be noted, that T_p and Curie temperature T_c of the films seem to differ strongly. Finally, the most pronounced hysteretic behavior is observed for the film of medium thickness h_F =69nm. In contrast, the weakest indication for hysteretic behavior is found in this analysis (fig. 3) for the thickest film. This might be taken as an additional indication for the effect of relaxation of the strain in thick STO films. A detailed discussion of the analysis of the ferroelectricity in these films will be given in a forthcoming paper. In this paper, we will concentrate on the technical aspects, i.e. the tunability and the multilayer properties.

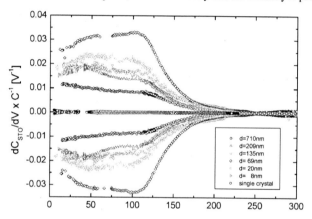

Fig. 3: Temperature dependence of the slope dC/dV at zero voltage obtained after application of large positive or negative bias fields, respectively, for STO films of different thickness and a STO single crystal.

Due to the effect of strain our films are ferroelectric up to 200-220K. As a consequence, all STO films show a large tunability up to elevated temperatures. Fig. 4 shows a comparison of the technical tunability of STO single and multilayers and a STO single crystal. The technical tunability is defined by:

$$n(E) = \frac{\varepsilon(E) - \varepsilon(0)}{\varepsilon(0)}. \tag{4}$$

The tunability of the single crystal drops exponentially with increasing temperature. It is below the limit of the measurement (10^{-3}) for T > 60K. The thin films show large tunability up to room temperature. At low temperatures, n ranges from n=1 to 5, at room temperature ten times smaller values (n=0.1 to 0.4) are measured. The multilayer shows a tunability that is similar to that of a STO single layer of thickness comparable to that of the individual STO layers.

All films show a tunability that might be technically relevant even for room temperature applications. The advantage of the multilayer is given by the fact, that the total capacitance is comparable to that of a thick STO film, whereas the properties of the individual layers are comparable to those of the thin STO films. Fig. 4 displays the temperature dependence of the capacitance and the inverse capacitance of 2 typical multilayers. The total capacitance of the multilayers ranges between 0.44 (at room temperature) and 0.9pF (at low temperature). A comparable single layer (e.g., with thickness $h_f = 25nm$) would produce an about 3 times smaller capacitance. Furthermore, the inverse capacitance shows a temperature dependence similar to that of the thin single layers, i.e., it is not Curie-Weiss like and shows a pronounced maximum at high temperatures.

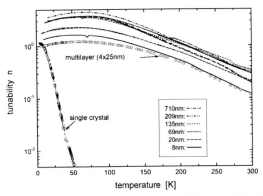

Fig. 4: Temperature dependence of the technical tunability for STO films of different thickness, a STO single crystal, and a multilayer with 4 layers STO (25nm) separated by 25nm thick CeO_2 layers. The bias field is $E=20V/\mu m$ for the films and 80V on the single crystal of 1mm thickness.

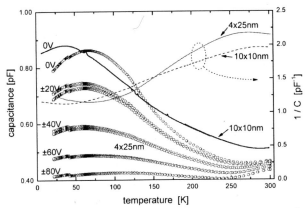

Fig. 4: Temperature dependence of the capacitance and inverse capacitance of two different multilayers consisting of ten 10nm thick STO layers and four 25nm thick STO layers separated by CeO_2 layers, respectively.

CONCLUSIONS

SrTiO$_x$ films with thicknesses ranging from 8nm to 710nm were fabricated on microwave suitable substrates, i.e., CeO_2 buffered sapphire substrates. The STO films grow epitaxially with (100) orientation with very similar structural properties on the (200) oriented CeO_2 buffer layer. The films are strained due to the lattice mismatch between the CeO_2 and due to the difference in thermal expansion of the sapphire substrate and the STO film. At room temperature, the c-axis length of the STO layers is only slightly elongated due to the lattice mismatch between CeO_2 and STO. The resulting tetragonality is estimated to be smaller than 0.3 %. The temperature dependence of the permittivity indicates, that the strain is relaxed over the STO film thicknesses $h_f > 100nm$, thinner films show a different temperature dependence and a strong enhancement of the permittivity at room temperature. Strain induced ferroelectricity persists up to $T_p = 200$-$220K$ resulting in a large tunability of n=1-5 at low temperature and n=0.1-0.4 at room temperature. Thus, biaxial compressive strain can lead to a considerable increase of the dielectric constant and tuning of $SrTiO_3$ thin films in technically relevant temperature regimes. Considering the advantage of the use of sapphire as a low-loss microwave material, this makes this system interesting for tunable microwave devices like varactors or phase shifters.

REFERENCES:
1. P. A. Fleury, J. M Worlock, Phys. Rev **174**, 613 (1968).
2. H. Vogt, Phys. Rev. B **51**, 8046 (1995).
3. K. A. Müller, R. Burkard, Phys Rev. B **19**, 3593 (1979).
4. R. Viana, P. Lunkenheimer, J. Hemberger, R. Böhmer, A. Loidl, Phys. Rev. B **50**, 601 (1994).
5. O. E. Kvyatkovskii, Solid State Commun. **117**, 455 (2001).
6. W. Zhong, D. Vanderbilt, Phys. Rev. B **53**, 5047 (1996).
7. J. G. Bednorz, K. A. Müller, Phys. Rev. Lett. **52**, 2289 (1984); W. Kleemann, J. Dec, Y. G. Wang, P. Lehnen, S. A. Prosandeev, J. Phys. Chem. Solids **61**, 167 (2000); C. Ang, Z. Yu, P. M. Vilarinho, J. L. Baptista, Phys. Rev. B **57**, 7403 (1998); W. Kleeman, A. Albertini, M. Kuss, R. Lindner, Ferroelectrics **203**, 57 (1997); W. Kleemann, J. Dec, Y. G. Wang, P. Lehnen, S. A. Prosandeev, J. Phys. Chem. Solids **61**, 167 (2000); G. A. Samara, J.Phys. Condens. Matter **15**, R367 (2003); B. E. Vugmeister M. D. Glinchuk, Rev. Mod. Phys. **62**, 993 (1990); C. Ang, Z. Yu, J. Appl. Phys. **91**, 1487 (2002); C. Ang, Z. Yu, Z. Jing, Phys. Rev. B **61**, 957 (2000).
8. E. Hegenbarth, Phys. Status Solidi **6**, 333 (1964); P. A. Fleury J. M. Worlock, Phys. Rev. **174**, 613 (1968); J. Hemberger, P. Lunkenheimer, R. Viana, R. Bohmer, A. Loidl, Phys. Rev. B **52**, 13159 (1995); D. Fuchs, C. W. Schneider, R. Schneider, H. Rietschel, J. Appl. Phys. **85**, 7362 (1999).
9. M. Itoh, R. Wang, Y. Inaguma, T. Yamaguchi, Y-J. Shan, T. Nakamura, Phys. Rev. Lett. **82**, 3540 (1999).
10. H. Uwe,T. Sakudo, Phys. Rev. B **13**, 271 (1976).
11. H.-C. Li, W. Si, R.-L. Wang, Y. Xuan, B. T. Liu, and X. X. Xi, Mater. Sci. Eng. B **56**, 218 (1998).
12. R. P. Lowndes, A. Rastogi, J. Phys. C **6**, 932 (1973).
13. N. A. Pertsev, A. K. Tagantsev, N. Setter, Phys. Rev. B **61**, R825 (2000); **65**, 219901(E) (2002).
14. J. H. Haeni, P. Irvin, W. Chang, R. Uecker, P. Reiche, Y. L. Li, S. Choudhury, W. Tian, M. E. Hawley, B. Craigo, A. K. Tagantsev, X. Q. Pan, S. K. Streiffer, L. Q. Chen, S. W. Kirchoefer, J. Levy, D. G. Schlom, Nature **430**, 758 (2004).
15. M. J. Dalberth, R. E. Stauber, J. C. Price, D. Galt, C. T. Rogers, Appl. Phys. Lett. **72**, 507 (1998).
16. R. Wördenweber, E. Hollmann, Mahmood Ali, J. Schubert, G. Pickartz, Tai Keong Lee, J. Europ. Ceramic Soc. (2006)
17. P. Kuzel, F. Kadlec, H. Nemec, R. Ott, E. Hoffmann, N. Klein, Appl. Phys. Lett. **88**, 102901 (2006)
18. O. G. Vendik, M. A. Nikol`skii, Techn. Phys. **46**, 112 (2001)

CONSTRUCTION AND CHARACTERIZATION OF (Y,Yb)MnO₃/HfO₂ STACKING LAYERS FOR APPLICATION TO FERAM

CONSTRUCTION AND CHARACTERIZATION OF (Y,Yb)MnO$_3$/HfO$_2$ STACKING LAYERS FOR APPLICATION TO FERAM

Kazuyuki Suzuki, Kaori Nishizawa, Takeshi Miki and Kazumi Kato
National Institute of Advanced Industrial Science and Technology (AIST)
2266-98 Anagahora, Shimoshidami, Moriyama-ku
Nagoya, 463-8560, Japan

ABSTRACT

$Y_{0.5}Yb_{0.5}MnO_3$ ferroelectrics/HfO$_2$ stacking layer were constructed on Si(100) substrates via chemical solution deposition. The HfO$_2$ insulating layer crystallized on Si(100) substrates and consisted of uniform grains and had a smooth surface. The $Y_{0.5}Yb_{0.5}MnO_3$ film prepared on the HfO$_2$ insulator layer had preferred orientation along the c-axis and consisted of uniform grains and a smooth surface. The degree of c-axis orientation and microstructure of the $Y_{0.5}Yb_{0.5}MnO_3$ films varied with the concentration of HfO$_2$ and $Y_{0.5}Yb_{0.5}MnO_3$ solutions. Also, the electrical properties of the $Y_{0.5}Yb_{0.5}MnO_3$ ferroelectrics/HfO$_2$ stacking layers depended on the concentration of the solutions. The improved $Y_{0.5}Yb_{0.5}MnO_3$ ferroelectrics/HfO$_2$ stacking layer showed the retention property for over 10^4s.

INTRODUCTION

YMnO₃-based material having the hexagonal structure exhibit ferroelectric properties and a single polarized direction along the c-axis. YMnO₃ has a low dielectric constant of 20 and a spontaneous polarization of 5.0 μC/cm^2 in the form of a single crystal.[1] In addition to its low dielectric constant, the chemical characteristic of being volatile-element-free is attractive for applications in field effect transistor (FET) -type ferroelectric memories. Ferroelectric random access memories (FeRAM) using polarization switching of the ferroelectrics is one of the nonvolatile memories. Particularly, FET-type ferroelectric memory has received great attention as next-generation memory because it has an advantage of high speed, low power consumption, reducing the memory cell size and nondestructive readout.[2-3] This type of memory uses the remanent polarization of the ferroelectric to control the surface conductivity of silicon. One promising structure is the metal/ferroelectrics/insulator/semiconductor (MFIS) structure which have been studied using several ferroelectrics and insulators. The MFIS structure has not been realized yet but was proposed a long time ago, because it is difficult to obtain ferroelectric/Si structure having good interface. In this structure, the ferroelectric film must have a small dielectric constant in order to apply sufficient voltage on the ferroelectric film for polarization switching.

Conventionally, the YMnO₃ films were synthesized by sputtering, pulse laser deposition (PLD), molecular beam epitaxy (MBE) and metalorganic chemical vapor deposition (MOCVD).[4-6] Also, some groups reported the synthesis of YMnO₃ films via chemical solutions.[7-8] In our previous work, the solid solution of YMnO₃ and YbMnO₃ has been studied for the purposes that increase the stability of the hexagonal structure and lower the processing temperature.[9-11] The synthesis of (Y,Yb)MnO₃ thin films using alkoxy-derived solutions and the effects of the crystallization conditions on the crystallographic properties have been investigated in detail. On the basis of fundamental studies, we were able to find that crystallization in Ar was inevitable for improvement of the electrical properties. Additionally, the ferroelectric properties of the

17

(Y,Yb)MnO$_3$ thin films with metal-ferroelectrics-metal (MFM) and MFIS structures were reported.[12-13]

In this paper, MFIS structures were constructed using Y$_{0.5}$Yb$_{0.5}$MnO$_3$ as ferroelectrics and HfO$_2$ as an insulator though the chemical solution process. The effects of the preparation conditions of the HfO$_2$ layer and Y$_{0.5}$Yb$_{0.5}$MnO$_3$ film on the crystallographic appearance and electrical properties of the MFIS structures are investigated.

EXPERIMENTAL PROCEDURE

HfO$_2$ insulating layers were prepared by spin coating alkoxy-derived solutions, which were prepared using hafnium iso-propoxide and ethylene glycol monomethyl ether (EGMME). The concentration was adjusted to 0.1 or 0.05 mol/L. The p-Si(100) substrate was chemically cleaned using HF acid and then soaked in EGMME and dried prior to coating. The films were deposited in two steps of 1000 rpm for 3 s and 3000 rpm for 30 s. The as-deposited thin film was dried at 150°C and calcined at 350°C in air. Then, the film was heated by rapid thermal annealing at 750°C for 10 min in O$_2$.

For the preparation of precursor solutions for (Y,Yb)MnO$_3$ films, yttrium iso-propoxide, ytterbium iso-propoxide and manganese iso-propoxide were selected as starting chemicals. EGMME was selected as a solvent. Since alkoxides are extremely sensitive to moisture, the entire procedure was conducted in dry nitrogen. Yttrium iso-propoxide was dissolved in EGMME. Manganese iso-propoxide was added to the solution with atomic ratios of (Y+Yb):Mn=1:1. The concentration of the solutions was adjusted to 0.2 mol/L or 0.1 mol/L for (Y,Yb)MnO$_3$ films. The solutions were then reacted at a reflux temperature of 124°C for 2 h. (Y,Yb)MnO$_3$ films were prepared on the insulating layers using the (Y,Yb)MnO$_3$ precursor solutions. The films were deposited, dried and calcined in the same manner as the insulating films. Then, the film was heated by rapid thermal annealing at 750°C for in Ar.

The Y$_{0.5}$Yb$_{0.5}$MnO$_3$ ferroelectrics/HfO$_2$ stacking layers were constructed using the HfO$_2$ and Y$_{0.5}$Yb$_{0.5}$MnO$_3$ solutions with different concentrations. The thickness of HfO$_2$ and Y$_{0.5}$Yb$_{0.5}$MnO$_3$ films were constant at 10nm and 200nm, respectively. The preparation conditions of Y$_{0.5}$Yb$_{0.5}$MnO$_3$/HfO$_2$ stacking layers were shown in Table I.

The crystalline phase of the films was identified by using X-ray diffraction (XRD) measurements. The surfaces of the films were observed by atomic force microscopy (AFM). Pt top electrodes with 150μm diameter were deposited on the surface of the ferroelectric films through a metal mask by electron beam evaporation method for measurements of electrical properties. The leakage current densities of the films were measured using an electrometer (Keithley 6517). The capacitance-voltage (C-V) characteristics were measured at 1 MHz using an impedance analyzer (Agilent 4294A).

Table I. Preparation conditions of (Y,Yb)MnO₃/HfO₂ stacking layers

	Concentration of HfO₂ solutions	Thickness of HfO₂	Concentration of (Y,Yb)MnO₃ solutions	Number of layers	Annealing time per each layer	Thickness of (Y,Yb)MnO₃
(a)	0.1mol/L	10nm	0.2mol/L	5	10min	200nm
(b)	0.05mol/L	10nm	0.2mol/L	5	10min	200nm
(c)	0.05mol/L	10nm	0.1mol/L	10	5min	200nm

RESULTS AND DISCUUSION

The HfO$_2$ insulating layers prepared using 0.1 and 0.05 mol/L solutions had monoclinic structure and showed (-1 1 1) preferred orientation. The HfO$_2$ layer prepared using 0.05mol/L solution had higher crystallinity. These insulating layers consisted of uniform grains and had smooth surfaces. Figure 1 shows XRD profiles of the Y$_{0.5}$Yb$_{0.5}$MnO$_3$ films prepared on the HfO$_2$/Si. Hexagonal Y$_{0.5}$Yb$_{0.5}$MnO$_3$ was crystallized on both insulating layers and no secondary phase was observed. It was found that the Y$_{0.5}$Yb$_{0.5}$MnO$_3$ films showed preferred orientation along the c-axis. This preferred orientation is advantageous for the applications because the polarization axis of (Y,Yb)MnO$_3$ is along the c-axis. The Y$_{0.5}$Yb$_{0.5}$MnO$_3$ films on the HfO$_2$ layer prepared using the relatively dilute 0.05mol/L solution has a higher degrees of c-axis orientation (Fig.1(b) and (c)). Furthermore, the degree of c-axis orientation was slightly improved by using the 0.1mol/L Y$_{0.5}$Yb$_{0.5}$MnO$_3$ solution (Fig.1(c)). The modification of the HfO$_2$ thin layer was more effective for controlling the orientation of the Y$_{0.5}$Yb$_{0.5}$MnO$_3$ films.

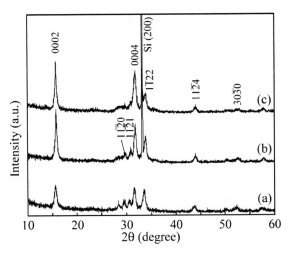

Figure 1 XRD profiles of Y$_{0.5}$Yb$_{0.5}$MnO$_3$/HfO$_2$ stacking layers crystallized on Si(100) substrates at 750°C via (a)0.1 and 0.2 mol/L, (b)0.05 and 0.2 mol/L and (c)0.05 and 0.1 mol/L precursor solutions for HfO$_2$ and Y$_{0.5}$Yb$_{0.5}$MnO$_3$ layer, respectively.

Figure 2 shows AFM images of the surfaces of the Y$_{0.5}$Yb$_{0.5}$MnO$_3$ films prepared on HfO$_2$/Si. These films consisted of uniform grains and had smooth surfaces. The grain size of the Y$_{0.5}$Yb$_{0.5}$MnO$_3$ films on HfO$_2$ layer prepared using 0.05mol/L HfO$_2$ solution and 0.2 mol/L Y$_{0.5}$Yb$_{0.5}$MnO$_3$ solutions was about 50 nm (Fig.2(b)). The other two films consisted of smaller grains (Fig.2(a) and (c)). The difference of the grain size was produced by not only the thickness of each coating layer of Y$_{0.5}$Yb$_{0.5}$MnO$_3$, but in the difference of the crystallinity of the HfO$_2$ layer. The surface roughness (RMS) of the Y$_{0.5}$Yb$_{0.5}$MnO$_3$ films prepared using the 0.2 mol/L solution were about 2.5 nm. On the other hand, the RMS value of the $Y_{0.5}Yb_{0.5}MnO_3$ films prepared using the 0.1 mol/L was about 3.5 nm and larger than that prepared using the 0.2 mol/L solution. The large surface roughness of the $Y_{0.5}Yb_{0.5}MnO_3$ films prepared using the 0.1 mol/L solution was considered to be due to the large number of coating cycles.

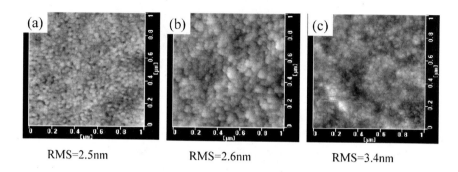

RMS=2.5nm RMS=2.6nm RMS=3.4nm

Figure 2 AFM images of $Y_{0.5}Yb_{0.5}MnO_3$/HfO₂ stacking layers crystallized on Si(100) substrates at 750°C via (a)0.1 and 0.2 mol/L, (b)0.05 and 0.2 mol/L and (c)0.05 and 0.1 mol/L precursor solutions for HfO₂ and $Y_{0.5}Yb_{0.5}MnO_3$ layer, respectively.

Figure 3 shows leakage current densities of the Pt/$Y_{0.5}Yb_{0.5}MnO_3$/HfO₂/Si structures. The leakage current densities of the Pt/$Y_{0.5}Yb_{0.5}MnO_3$/HfO₂/Si structures prepared by conditions (a), (b) and (c) (shown in table I) were 1.5×10^{-6}, 5.9×10^{-8} and 5.3×10^{-7} A/cm² at 5V, respectively. The large leakage current of the Pt/$Y_{0.5}Yb_{0.5}MnO_3$/HfO₂/Si structure prepared using 0.1 mol/L HfO₂ solution was due to the low crystallinity of the HfO₂ layer (Fig.3 (a)). Also, the large leakage current of the Pt/$Y_{0.5}Yb_{0.5}MnO_3$/HfO₂/Si structure prepared using 0.1 mol/L $Y_{0.5}Yb_{0.5}MnO_3$ solution (Fig.3 (c)) is associated with surface roughness and coarse grain boundaries, which work as a leak current path.

Figure 4 shows the C-V characteristics of the Pt/$Y_{0.5}Yb_{0.5}MnO_3$/HfO₂/Si structures measured at 1 MHz with a sweep rate of 0.2 V/s from -10 V to +10 V and vice versa. The clockwise C-V hysteresis loops induced by ferroelectric polarization switching were observed. The $Y_{0.5}Yb_{0.5}MnO_3$ on the HfO₂ layer prepared using 0.05mol/L solution showed a larger memory window (Fig.4 (b)and (c)). The electrical properties depended on the crystallinity and orientation of the $Y_{0.5}Yb_{0.5}MnO_3$ ferroelectric film, and the sharpness of the interface between the $Y_{0.5}Yb_{0.5}MnO_3$ ferroelectric film and HfO₂ insulating layer. It is important to control the surface structure of the HfO₂ layer in order to obtain advantageous crystallization of the $Y_{0.5}Yb_{0.5}MnO_3$ ferroelectric film and a sharp uniform interface.

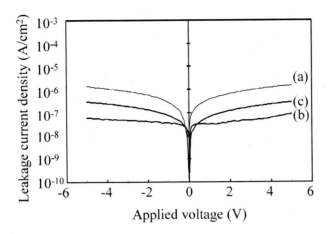

Figure 3 Leakage current density of Pt/Y$_{0.5}$Yb$_{0.5}$MnO$_3$/HfO$_2$/Si structures prepared via (a)0.1 and 0.2 mol/L, (b)0.05 and 0.2 mol/L and (c)0.05 and 0.1 mol/L precursor solutions for HfO$_2$ and Y$_{0.5}$Yb$_{0.5}$MnO$_3$ layer, respectively.

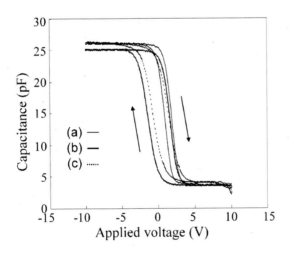

Figure 4 C-V characteristics of Pt/Y$_{0.5}$Yb$_{0.5}$MnO$_3$/HfO$_2$/Si structures prepared via (a)0.1 and 0.2 mol/L, (b)0.05 and 0.2 mol/L and (c)0.05 and 0.1 mol/L precursor solutions for HfO$_2$ and Y$_{0.5}$Yb$_{0.5}$MnO$_3$ layer, respectively.

Figure 5 shows retention properties at the write voltage of +10V or -10V for the Pt/Y$_{0.5}$Yb$_{0.5}$MnO$_3$/HfO$_2$/Si structures. In the case of the Pt/Y$_{0.5}$Yb$_{0.5}$MnO$_3$/HfO$_2$/Si structures with HfO$_2$ layer prepared using 0.1mol/L, the capacitance changed with time and the margin of the capacitance after 10^4 s decreased below half of the initial value (Fig.5 (a)). In contrast, the Pt/Y$_{0.5}$Yb$_{0.5}$MnO$_3$/HfO$_2$/Si structures using HfO$_2$ layer prepared using 0.05mol/L the capacitances were almost constant over 10^4s.

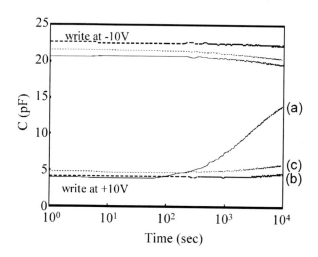

Figure 5 Retention properties of Pt/Y$_{0.5}$Yb$_{0.5}$MnO$_3$/HfO$_2$/Si structures prepared via (a)0.1 and 0.2 mol/L, (b)0.05 and 0.2 mol/L and (c)0.05 and 0.1 mol/L precursor solutions for HfO$_2$ and Y$_{0.5}$Yb$_{0.5}$MnO$_3$ layer, respectively.

The preparation conditions of the HfO$_2$ layer was found to be important for controlling the crystallographic appearance and electrical properties of the MFIS structures. For downsizing of the Pt/Y$_{0.5}$Yb$_{0.5}$MnO$_3$/HfO$_2$/Si structures, modification of the HfO$_2$ layer is important to obtain a sharp interface structure and to control the electrical properties.

CONCLUSION

Y$_{0.5}$Yb$_{0.5}$MnO$_3$ (200nm)/HfO$_2$ (10nm)/Si staking layers were constructed by chemical solution deposition. The degree of c-axis orientation and microstructure of Y$_{0.5}$Yb$_{0.5}$MnO$_3$ films depended on the concentration of the HfO$_2$ and Y$_{0.5}$Yb$_{0.5}$MnO$_3$ solutions. Furthermore, the electrical properties varied with the concentration of solutions and the resultant structures. The modification of the HfO$_2$ layer was more effective for the improvement of the crystallographic orientation, microstructure and electrical properties of the Y$_{0.5}$Yb$_{0.5}$MnO$_3$ ferroelectrics/HfO$_2$ stacking layer. The improved Y$_{0.5}$Yb$_{0.5}$MnO$_3$ ferroelectrics/HfO$_2$ stacking layer showed the retention property over 10^4s.

REFERENCES

[1]G.A. Smolenskii and V.A. Bokov, "Coexistence of magnetic and electric ordering in crystals", *J. Appl. Phys.*, **35**, 915-18 (1964).

[2]T. Hirai, Y. Fujisaki, K. Nagashima, H. Koike and Y. Tarui, "Preparation of SrBi$_2$Ta$_2$O$_9$ Film at Low Temperatures and Fabrication of a Metal/Ferroelectric/Insulator/Semiconductor Field Effect Transistor Using Al/SrBi$_2$Ta$_2$O$_9$/CeO$_2$/Si(100) Structures", *Jan. J. Appl. Phys.*, **36**, 5908-11 (1997).

[3]E. Tokumitsu, G. Fujii and H. Ishiwara, "Nonvolatile ferroelectric-gate field-effect transistors using SrBi$_2$Ta$_2$O$_9$/Pt/SrTa$_2$O$_6$/SiON/Si structures", *Appl. Phys. Lett.*, **75**, 575-77 (1999).

[4]H.N. Lee, Y.T. Kim and S.H. Choh, "Comparison of memory effect between YMnO$_3$ and SrBi$_2$Ta$_2$O$_3$ ferroelectric thin films deposited on Si substrates", *Appl. Phys. Lett.*, **76**, 1066-1068 (2000).

[5]N. Fujimura, T. Ishida, T. Yoshimura and T. Ito, "Epitaxially grown YMnO$_3$ film: New candidate for nonvolatile memory devices", *Appl. Phys. Lett.*, **69**, 1011-1013 (1996).

[6]S. Imada, T. Kuraoka, E. Tokumitsu and H. Ishiwara, " Ferroelectricity of YMnO$_3$ thin films on Pt(111)/Al$_2$O$_3$(0001) and Pt(111)/Y$_2$O$_3$(111)/Si(111) structures grown by molecular beam epitaxy", *Jpn. J. Appl. Phys.* **40**, 666-671 (2001).

[7]N. Fujimura, H. Tanaka, H. Kitahata, K. Tadanaga, T. Yoshimura, T. Ito and T.Minami, "YMnO$_3$ thin films prepared from solutions for non volatile memory devices", *Jan. J. Appl. Phys.*, **36**, L1601-L1603 (1997).

[8]W.C. Yi, J.S. Choe, C.R. Moon, S.I. Kwun and J.G. Yoon, "Ferroelectric characterization of highly (0001)-oriented YMnO$_3$ thin films grown by chemical solution deposition", *Appl. Phys. Lett.*, **73**, 903-905 (1998).

[9]K. Suzuki, K. Nishizawa, T. Miki and K. Kato, " Synthesis of ferroelectric YMnO$_3$ thin film by chemical solution deposition" *Key. Eng. Mater.*, **214-215**, 151-156 (2002).

[10]K. Suzuki, K. Nishizawa, T. Miki and K. Kato, " Effects of annealing conditions on crystallization of hexagonal manganite films", *Ferroelectrics.*, **270**, 99-104 (2002).

[11]K. Suzuki, D. Fu, K. Nishizawa, T. Miki and K. Kato, "Effects of substrates on alkoxy-derived (Y,Yb)MnO$_3$ thin films", *Integ. Ferroelectr.*, **47**, 91-100 (2002).

[12]K. Suzuki, D. Fu, K. Nishizawa, T. Miki and K. Kato, "Preparation of (Y,Yb)MnO$_3$/Y$_2$O$_3$/Si (MFIS) structure by chemical solution deposition method", *Jpn. J. Appl. Phys.*, **42**, 6007-6010 (2003).

[13]K. Suzuki, D. Fu, K. Nishizawa, T. Miki and K. Kato, "Ferroelectric properties of (Y,Yb)MnO$_3$ thin films prepared using alkoxide solutions", *Key. Eng. Mater.*, **248**, 77-80 (2003).

TEMPERATURE DEPENDENCE ON THE STRUCTURE AND PROPERTY OF Li$_{0.06}$(Na$_{0.5}$K$_{0.5}$)$_{0.94}$NbO$_3$ PIEZOCERAMICS

Ken-ichi Kakimoto, Kazuhiko Higashide, Tatsuro Hotta, and Hitoshi Ohsato
Department of Materials Science and Engineering, Graduate School of Engineering,
Nagoya Institute of Technology
Gokiso-cho, Showa-ku, Nagoya 466-8555, Japan

ABSTRACT

Lead-free (Na$_{0.5}$K$_{0.5}$)NbO$_3$ (NKN) - LiNbO$_3$ (LN) mixture system is now studied most intensively for piezoelectric transducer application. In this study, the temperature dependence on the piezoelectric property of Li$_{0.06}$(Na$_{0.5}$K$_{0.5}$)$_{0.94}$NbO$_3$ ceramics was evaluated. The variation in the property was discussed in the lattice deformation characterized by Raman scattering measurement at various temperatures. The internal vibration modes of the NbO$_6$ octahedron were correlated well with the phase transition and the property change.

INTRODUCTION

Li$_x$(Na$_{0.5}$K$_{0.5}$)$_{1-x}$NbO$_3$ (LNKN) ceramics with high Curie temperatures above 400°C is now studied most intensively from the practical point of view for lead-free piezoelectric transducer application needed for operating at wide temperatures. In particular, we reported that LNKN ceramics with x=0.06, Li$_{0.06}$(Na$_{0.5}$K$_{0.5}$)$_{0.94}$NbO$_3$, show the maximized best properties of d_{33}=235 pC/N, k_p = 0.42 and k_t = 0.48 at room temperature.[1] The origin of these good properties has been considered to correlate with a kind of "morphotropic-phase-boundary (MPB) effect" that a phase boundary between orthorhombic and tetragonal perovskites appears at temperature close to the room temperature. In fact, a dielectric anomaly observed at temperature around 200°C for (Na$_{0.5}$K$_{0.5}$)$_{1-x}$NbO$_3$ (NKN) ceramics results from the phase transition is shifted to lower temperature toward the room temperature as increasing Li content. However, we have as yet no information on the phase transition of Li$_{0.06}$(Na$_{0.5}$K$_{0.5}$)$_{0.94}$NbO$_3$ ceramics and its exact relation to the piezoelectric property.

In this study, the phase transition of Li$_{0.06}$(Na$_{0.5}$K$_{0.5}$)$_{0.94}$NbO$_3$ ceramics was evaluated by temperature-variable X-ray diffraction and Raman scattering measurements, then the results were compared with the piezoelectric property measured at various temperatures in order to characterize the effect of the phase transition on the property.

EXPERIMENTAL PROCEDURE

Li$_{0.06}$(Na$_{0.5}$K$_{0.5}$)$_{0.94}$NbO$_3$ ceramics was prepared by the mixed oxide method using high-purity Li$_2$CO$_3$, Na$_2$CO$_3$, K$_2$CO$_3$, and Nb$_2$O$_5$ powders as starting materials. The oxide powders were mixed, and ball-milled for 24 h using ethanol as a medium. A dried sample was calcinated at 850°C. After crushing and adding polyvinyl alcohol, the granulated powders were pressed into cylindrical forms, followed by sintering in air at 1082°C.

The crystal structure of Li$_{0.06}$(Na$_{0.5}$K$_{0.5}$)$_{0.94}$NbO$_3$ ceramics at temperatures above the room temperature was determined by X-ray powder diffraction (XRPD) using a Philips X-Pert Diffractometer (Cu$K\alpha$) equipped with a sample heating holder (Pt heater). The Raman scattering spectra were excited using 514.5 nm radiation from an Ar$^+$ laser and were collected by a microscopic Raman spectrometer (JASCO, NRS-2000) in the 45-1062 cm^{-1} range at various temperatures ranging from -150°C to 450°C using a sample cooling/heating unit.

For electric measurement, silver paste was fired on both surfaces of the ceramic specimens as electrodes. Their dielectric constants at 1 kHz as a function of temperature were measured using a LCR meter. Samples for the piezoelectric measurements were poled at 150°C in a silicon oil bath by applying a dc electric field of 3 kV/mm. The temperature dependence of the planar electromechanical coupling factor (k_p) was determined from the resonance-antiresonance method using an Agilent 4294A impedance analyzer and an environmental testing chamber. Piezoelectric d_{33} constants were measured using the quasi-static method by a Berlincourt-type d_{33} meter.

RESULTS AND DISCUSSION

Figure 1 shows the temperature dependence of the dielectric constant at 1 kHz for $Li_{0.06}(Na_{0.5}K_{0.5})_{0.94}NbO_3$ ceramics. The temperature of the maximum dielectric constant (T_{max}) appears at 442°C, and this temperature region corresponds to the transition between tetragonal (ferroelectric) and cubic (paraelectric) phases, namely the Curie point. On the other hand, the above-mentioned dielectric anomaly is observed at the temperature ranging from -40 to 150°C, and its peak-top is located at 14°C.

The XRPD patterns measured at the temperature from 30 to 120°C in 10°C interval for $Li_{0.06}(Na_{0.5}K_{0.5})_{0.94}NbO_3$ ceramics are presented in Figure 2. This temperature range is selected inside the temperature range showing the dielectric anomaly. The patterns are almost identical in shape and intensity despite with different measurement temperatures, and all of them can be judged to be a tetragonal perovskite structure by assuming the formation of a single phase. However, since the doping amount of Li is slight 6 at% to NKN, it is difficult to be considered that all crystal structure in the material forms a uniform tetragonal perovskite, and it may be the mixed crystal that microscopically consists of tetragonal and orthorhombic phases.

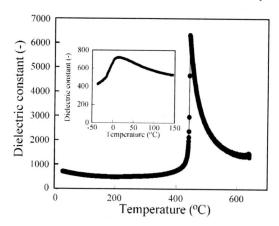

Figure 1. Temperature dependence of dielectric constant at 1 kHz for $Li_{0.06}(Na_{0.5}K_{0.5})_{0.94}NbO_3$ ceramics.

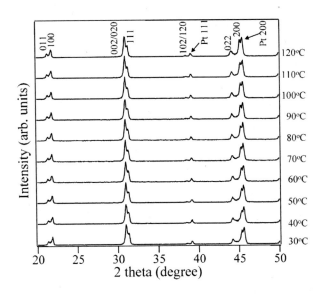

Figure 2. XRPD patterns of $Li_{0.06}(Na_{0.5}K_{0.5})_{0.94}NbO_3$ ceramics.

XRPD allows the confirmation of macroscopic symmetry in long-range ordering, but is not very sensitive to nonuniform distortions of the crystal lattice in short-range ordering. Therefore, the Raman scattering technique was applied to describe the detailed structural change of $Li_{0.06}(Na_{0.5}K_{0.5})_{0.94}NbO_3$ ceramics against temperature. Figure 3 shows the Raman spectra of $Li_{0.06}(Na_{0.5}K_{0.5})_{0.94}NbO_3$ ceramics at various temperature ranging from -150 to 500°C. We have already reported that the vibrations recorded in $Li_x(Na_{0.5}K_{0.5})_{1-x}NbO_3$ ceramics can be separated to lattice translations involving motions of the alkaline cations and internal modes of NbO_6 octahedra, and further the vibrations of the NbO_6 octahedron can be classified as $1A_{1g}$ (v_1) + $1E_g$ (v_2) + $2F_{1u}$ (v_3, v_4)+ F_{2g} (v_5) + F_{2u} (v_6).[2] These assignments are described in this figure as notations.

The translational modes of Na^+/K^+ cations are well observed at temperatures below 0°C, but start to submerge under a strong elastic scattering coming from zone center at around 0°C, and finally it is not possible to observe them clearly over 25°C. In contrast, stretching v_1 and bending v_5 modes are detected as relatively strong scatterings at all the temperatures investigated. The stretching v_1 mode represents an equilateral symmetry of NbO_6 octahedron, and the peak shifts to a lower frequency with increasing temperature on the whole, as shown in Figure 4, is due to a decrease in binding strength caused by the extending of the distance between niobium and its coordinate oxygen, mainly resulting from the thermal expansion. However, this figure also demonstrates that the peak shift has a discontinuity at temperatures between 0 and 100°C. In addition, it is noted that the scattering intensity of v_2, which is observed in the left side of v_1 peak as a shoulder hump in the Raman spectrum, weakens as increase of temperature from 0°C, and finally the hump becomes the uncertainty beyond 100°C.

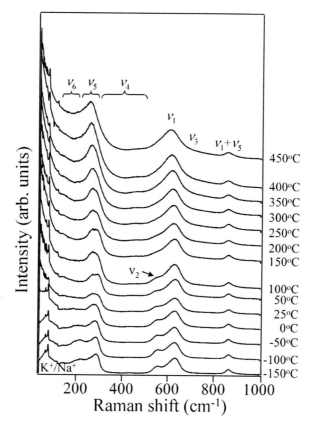

Figure 3. Raman scattering spectra of $Li_{0.06}(Na_{0.5}K_{0.5})_{0.94}NbO_3$ ceramics.

These facts clearly show that the symmetry of the NbO_6 octahedron changes in short-range ordering at temperatures ranging from 0 to 100°C, and it is estimated that $Li_{0.06}(Na_{0.5}K_{0.5})_{0.94}NbO_3$ ceramics started its phase transformation, orthorhombic-tetragonal transition, in this temperature range. This phase transformation seems to have occurred locally and gradually against temperature change, based on the distribution characteristic of Li that dissolves in NKN, and this resulted in an observed dielectric anomaly in the relatively wide temperature range (Figure 1). In this temperature range, a characteristic change also appears in v_5 simultaneously. The peak shape of v_5 changes clearly and the tip becomes broad with the increasing temperature close to 150°C, but it sharpens again in the higher temperatures above 200°C. In addition, the peak shifting to lower wavenumber caused by the effect of the thermal expansion is also observed similar to the case of v_1.

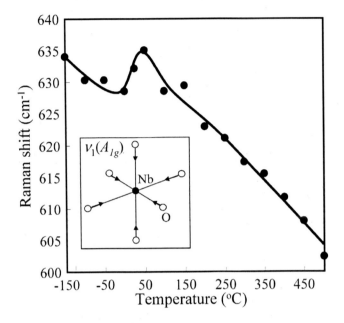

Figure 4. Raman shift of stretching v_1 mode.

Figure 5 presents the temperature dependence on the piezoelectric property (k_p and d_{33}) for $Li_{0.06}(Na_{0.5}K_{0.5})_{0.94}NbO_3$ ceramics. When k_p was measured at temperatures from -40 to 70°C, the enhanced k_p was observed in the range from 0 to 40°C as a broad hump centered at 20°C. In this temperature range, dielectric anomaly was observed and the peak shape of each Raman scattering mode changes, as mentioned above. In other words, the piezoelectric property is closely related to the phase transformation. Furthermore, it was observed from the measurement of frequency constant that the elastic compliance as well as dielectric constant for $Li_{0.06}(Na_{0.5}K_{0.5})_{0.94}NbO_3$ ceramics rapidly changes in this temperature range.[3]

On the other hand, another temperature dependence is observed in the d_{33} measurement. The enhanced property, d_{33}=235 pC/N, is maintained to around 200°C that is higher than the phase transition temperature, then turns to a plateau in the d_{33} value of around 215 pC/N up to near the Curie temperature. It is empirically accepted that d_{33} value is a structure-sensitive physical property compared to k_p. Therefore, it is presumed that d_{33} more reflects the local structural change. The temperature dependence of d_{33} resembles especially the temperature change observed in the bending v_5 mode rather than the stretching v_1 mode. In the previous study,[2] we reported that v_5 mode well exhibits a difference in the distorted NbO_6 polyhedral configuration between $(Na,K)NbO_3$ perovskite and $LiNbO_3$ ilmenite units. Therefore, more detailed structural analysis in short-range ordering besides the phase transition, e.g., lattice distortion at local micro area, should be carried out to investigate the relation to the d_{33} variation.

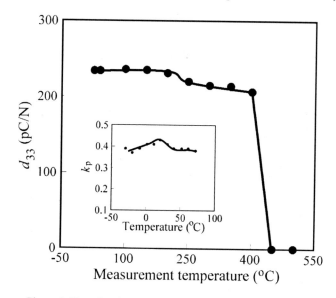

Figure 5. Piezoelectric property of $Li_{0.06}(Na_{0.5}K_{0.5})_{0.94}NbO_3$ ceramics.

CONCLUSION

In conclusion, temperature dependence was observed in the piezoelectric property of $Li_{0.06}(Na_{0.5}K_{0.5})_{0.94}NbO_3$ ceramics, and the enhanced piezoelectric property at room temperature is mainly related to the existence of the phase transformation. However, their good piezoelectric property is not limited in the narrow temperature regions, but kept in a wide operating temperature close to the Curie temperature above 400°C.

ACKNOWLEDGEMENT

This work was supported by a Grant-in-Aid for Scientific Research for Encouragement of Young Scientists (B) (No. 17760544) form the Japan Society for the Promotion of Science (JSPS), and by research grants from The Matsuda Foundation (2004), The Kazuchika Okura Memorial Foundation (2005), Research Foundation for the Electrotechnology of Chubu (2006) and from the NITECH 21st Century COE Program "World Ceramics Center for Environmental Harmony".

REFERENCES
[1]Y. Guo, K. Kakimoto, and H. Ohsato, "Phase Transitional Behavior and Piezoelectric Properties of $(Na_{0.5}K_{0.5})NbO_3$-$LiNbO_3$ Ceramics," *Appl. Phys. Lett.*, **85**, 4121-23 (2004).
[2]K. Kakimoto, K. Akao, Y. Guo and H. Ohsato, "Raman Scattering Study of Piezoelectric $(Na_{0.5}K_{0.5})NbO_3$-$LiNbO_3$ Ceramics," *Jpn. J. Appl. Phys.*, **44**, 7064-67 (2005).
[3]K. Higashide, K. Kakimoto, and H. Ohsato, "Temperature Dependence on the Piezoelectric Property of $(Na_{0.5}K_{0.5})NbO_3$-$LiNbO_3$ Ceramics," *J. Eur. Ceram. Soc.*, *submitted*.

POLAR AXIS ORIENTATION AND ELECTRICAL PROPERTIES OF ALKOXY-DERIVED ONE MICRO-METER-THICK FERRO-/PIEZOELECTRIC FILMS

Kazumi Kato[1,2], Shingo Kayukawa[2], Kazuyuki Suzuki[1], Yoshitake Masuda[1], Tatsuo Kimura[1], Kaori Nishizawa[1], Takeshi Miki[1]

[1]National Institute of Advanced Industrial Science and Technology, 2266-98 Anagahora, Shimoshidami, Moriyama-ku, Nagoya 463-8560, Japan

[2]Nagoya Institute of Technology, Gokiso, Showa-ku, Nagoya 466-8555, Japan

ABSTRACT

Polar-axis oriented $CaBi_4Ti_4O_{15}$ (CBTi144) films were fabricated on Pt foils using a complex metal alkoxide solution. Ambient during heat treatments such as pre-baking and crystallization affected the nucleation behaviors of the alkoxy-derived films. It was found that the pre-baking ambient affected the microstructure, and the crystallization ambient impacted the crystal perfection and the ferroelectric properties. The 500 nm-thick film that was pre-baked in air and then crystallized in oxygen flow consisted of columnar grains with rather rough surface and showed good ferroelectric properties. Closely-packed dense structure with relatively smooth surface was obtained by both pre-baking and crystallization in oxygen flow. The 1 μm-thick CBTi144 films exhibited low leakage current density of 10^{-7} A/cm^2 and high piezoelectric constant, d_{33} of 260 pm/V.

INTRODUCTION

$CaBi_4Ti_4O_{15}$ (CBTi144) is a member of the Aurivillius family, in which the pseudo-perovskite layers and bismuth-oxygen layers are stacking alternatively along c axis[1]. CBTi144 with even numbers of stacking oxygen octahedron along c axis have no polarization along c axis but spontaneous polarization along a axis, because there is a mirror plane perpendicular to the c axis. Therefore, the thin films with a axis orientation are preferred for use in many kind of devices such as ferroelectric random access memories (FeRAM), piezoelectric microactuators and resonators. Additionally, CBTi144 is characterized as its high Currie point of about 790°C[2], therefore is expected for special applications at relatively high temperature.

In our previous work[3], the polar-axis oriented CBTi144 thin films have been fabricated via the chemical solution deposition (CSD) technique using a complex alkoxide solution directly on both sides of Pt foils showing preferred orientation of (100) and (110). The polar-axis orientation of CBTi144 thin films was strongly associated with the preferred orientation of the Pt foils and stems from the good lattice matching between the c axis of CBTi144 and the 110 direction in the Pt(100) plane. The ferroelectric and piezoelectric properties of the polar-axis oriented CBTi144 thin film

33

had been improved compared with the randomly-crystallized thin films on Pt-coated Si[4-7]. The enhanced properties were considered to associate with high degrees of the polar-axis orientation and good uniformity of the effectively-applied electric field within the film. Recently, impacts of oxygen ambient during crystallization of the polar-axis oriented CBTi144 films on the crystal quality and electrical properties have been preliminary investigated[8-10]. It was found that the ferroelectric and piezoelectric properties were improved by crystallization in oxygen flow. The improvement seems to associate with oxygen stoichiometry. In this paper, relations of ambient during pre-baking and crystallization of the alkoxy-derived films and the crystallinity, microstructure and electrical properties are investigated. The nucleation behaviors in the complex alkoxy-derived films are discussed. Finally, the structural and electrical properties of CBTi144 thick films are improved.

EXPERIMENTAL

A precursor solution of Ca-Bi-Ti complex alkoxide for CBTi144 films was prepared through chemical reactions of Ca metal, Bi triethoxide, Ti isopropoxide in a mixture of ethanol and 2-methoxyethanol[11]. The chemical composition of Ca:Bi:Ti is adjusted to stoichiometric 1:4:4. The films were deposited on both sides of Pt foils with a thickness of 0.020 mm by dip-coating method. The foils were immersed in the precursor solution and were perpendicularly withdrawn upward at a constant speed of 1.5 mm/s in air. The as-deposited films were dried at 130°C in air and followed by pre-baking at 350°C in air or oxygen flow. Finally, the films were crystallized at 700°C for 10 min in air or oxygen flow. The purity of oxygen was 99.5 %. The oxygen ambient was controlled by using a flow meter. The oxygen partial pressure was given as almost 1 atm at a flow rate of 2 L/min. The film thickness was adjusted by multiplication of the dip-coating and heat treatments. The crystal phase and crystallographic orientation of the films were identified using X-ray diffraction (XRD) measurements using Cu Kα radiation. The acceleration voltage and current were 40 kV and 40 mA, respectively. Microstructure of the films was observed using a field emission scanning electron microscopy (FESEM) at acceleration voltage of 15 kV and a high resolution transmission electron microscopy (HRTEM) at acceleration voltage of 300 kV. Prior to electrical measurements, circular Pt electrodes of 150 μm diameter were deposited through a metal mask by electron beam evaporation method. Leakage current densities were measured by an electrometer. Dielectric and ferroelectric properties were measured using an impedance analyzer and a ferroelectric test system at room temperature, respectively. The piezoelectric properties were evaluated using a piezoelectric force microscopy (PFM).

RESULTS AND DISCUSSION

Figure 1 shows XRD profiles of CBTi144 films heat treated in various ambient. CBTi144 films contained no sub-phase such as pyrochlore which often dominantly crystallized[12], and showed high intensities of (200)/(020) diffraction line compared to the other lines, although the (200) and (020) diffraction lines could not be distinguished. The crystallographic characteristic is considered to be due to good matching between the c-axis of CBTi144 film and the 110 direction in the (100) plane of Pt foil[3,13]. The (200)/(020) diffraction lines for the CBTi144 films crystallized at 700°C in oxygen flow were found to be narrower and sharpened compared with that of the film crystallized in air. The narrowness and sharpness of the diffraction line indicates that degree of the crystal perfection of the film crystallized in oxygen flow is much higher. As organic residue in the alkoxy-derived films oxidize to gas phase and remove, it is necessary for the films to be heated in excess oxygen in order to maintain the oxygen stoichiometry. The crystal perfection may associate with oxygen stoichiometry of the films. Figure 2 shows FE-SEM cross-section profiles of the films heat treated in various ambient. The thickness of the films was about 500 nm or a little thicker. It appeared that the films were divided into two categories; columnar grain with rough surface and closely-packed dense structure with relatively flat surface. The films pre-baked in air had the former structure and the films pre-baked in oxygen flow had the latter structure. The pre-baking ambient had significant effects on the microstructure.

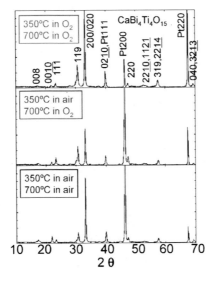

Figure 1   XRD profiles of CBTi144 films heat treated in various ambient[10].

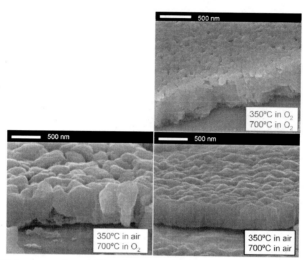

Figure 2   FE-SEM cross section profiles of CBTi144 films heat treated in various ambient[10].

In the case of pre-baking in air, the elimination of gas phase and nucleation proceeded simultaneously in the films, so that the homogeneous nucleation inner the film is predominant and resultantly the columnar grains developed after multiple deposition.   In contrast, the film is carbon-free and pure amorphous Ca-Bi-Ti-O in the case of pre-baking in oxygen flow, the nucleation is predominant at the surface of the platinum where free energy for crystallization is minimum because of the similar atomic arrangement.   The closely-packed dense structure with relatively smooth surface and high orientation degrees maintained even after huge multiple deposition.

Figure 3 shows I-V characteristics of CBTi144 films heat treated in various ambient.   The film heat treated in oxygen flow exhibited the lowest leakage current density.   For the improved CBTi144 films, the applied voltage would effectively work for the polarization switching.   The dielectric constant and loss factor of the CBTi144 films crystallized in oxygen were 289 and 0.08, respectively, at a frequency of 1 kHz.   The dielectric constant changed gradually with frequency of 100Hz to 1 MHz.   Figure 4 shows P-V characteristics of Pt/CBTi144(thickness: 500nm)/Pt capacitors at an applied voltage of 50 V and a frequency of 500 Hz.   The remanent polarization (Pr) and coercive electric field (Ec) of the CBTi144 film crystallized in oxygen flow were 33.6 $\mu C/cm^2$ and 357 kV/cm, respectively.   For comparison, the P-V hysteresis loop of the film crystallized in air was plotted.   The Pr and Ec were 17.0 $\mu C/cm^2$ and 225 kV/cm, respectively.

Also, it should be noticed that the Pr and Ec values were enhanced more than twice of the values obtained for CBTi144 thin films with random orientation[4,5]. Figure 5 shows the voltage dependence of the Pr and Ec for the 500 nm-thick CBTi144 films crystallized in oxygen flow and air. The Pr and Ec increased with applied voltage. The changes of the Pr and Ec values and the saturation characteristics were entirely distinguished for the both films. The Pr and Ec increased at up to 50 V and saturated thereafter with respect to the film crystallized in oxygen. Much higher voltage was necessary for full polarization switching to be applied to the film crystallized in air. At the applied voltage of 115 V, the Pr and Ec took maximum as 24.7 $\mu C/cm^2$ and 306 kV/cm, respectively. It was found that the voltage for switching polarization was able to be lowered as less than half by controlling oxygen stoichiometry of the CBTi144 films. Figure 6 shows the PFM experimental set-up and d_{33}-V hysteresis loop of the CBTi144 films crystallized in oxygen flow. In order to apply the electric field homogenously to the film and measure the displacement precisely, the conductive W_2C-coated Si cantilever tip was contacted on the Pt top electrode. A force constant and a mechanical resonance frequency of the conductive cantilever were 7.0 N/m and 180 kHz, respectively, and have been confirmed to be proper for the piezoelectric measurements. As the diameter of the conductive cantilever tip was about 30 nm, the physical and electrical contacts were considered to be sufficient for the measurements. In the d_{33} measurements, the film was first polarized by an 100 ms pulse at various voltages, then measurement was performed within the following 500 ms. As on-top electrode measurements, the depolarization field has been completely compensated, and the observed value represents the quasistatic state of the film. Therefore, no relaxation was observed in the d_{33} loop measurement when lowering the poling voltage from maximum to zero. The d_{33} values were 140 pm/V at maximum poling voltage of 60 V and 260 pm/V at 30 V for the 500 nm-thick CBTi144 films crystallized in air and oxygen flow, respectively.

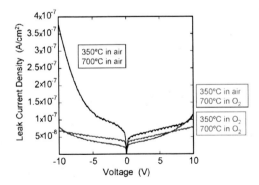

Figure 3   I-V characteristics of CBTi144 films heat treated in various ambient[10].

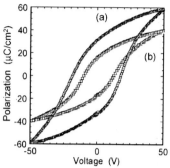

Figure 4   P-V characteristics for 500 nm-thick CaBi₄Ti₄O₁₅ films crystallized on Pt foils at 700°C in (a) oxygen flow and (b) air[9].

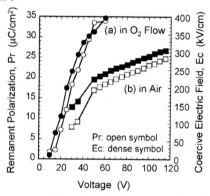

Figure 5   Voltage dependences of the Pr and Ec for the CaBi₄Ti₄O₁₅ films crystallized in (a) oxygen flow and (b) air[8].

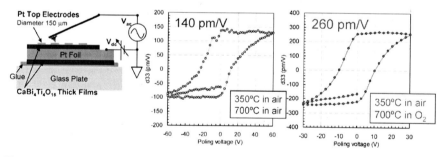

Figure 6 shows the PFM experimental set-up and d_{33}-V hysteresis loop of the 500 nm-thick CBTi144 films crystallized in oxygen flow.

Figure 7 shows cross-section TEM profile of 1 μm-thick CBTi144 film and electron diffraction patterns of the selected area. It was found that the film pre-baked in oxygen flow and then crystallized in oxygen flow was completely densified along both of the in-plane and out-of plane directions. There have been no crack and void in the film. The structure is considered to be based on controlled nucleation at the bottom sites close to the interface and the gradual growth to the upper in each deposition layer. The electron diffraction patterns indicated that the film exhibited (100)/(010) orientation. The result is in good agreement with the XRD data as shown Fig. 1. Figure 8 shows the topography of the Pt top electrode for the PFM measurements and d_{33}-V characteristic of 1 μm-thick CBTi144 film on Pt foil. The surface of the Pt top electrode, which was affected by the surface of CBTi144 film, was relatively flat; the RMS value was about 23.5 nm. The d_{33} at the maximum voltage of 60 V was about 260 pm/V, which is identical to the value of the 500 nm-thick CBTi144 film. It indicates that the polar axis oriented CBTi144 thick film completely responded to the applied voltage. The domains reversed and then got strained when further lowering the poling voltage zero to minimum. The absolute strain increased to at the minimum poling voltage of -60 V. The asymmetric appearance of the d_{33}-V curve is considered to associate with the difference of the top and bottom electrode sizes. The enhancement of the d_{33} is considered to be due to the higher degree of the polar axis orientation and the closely-packed dense structure.

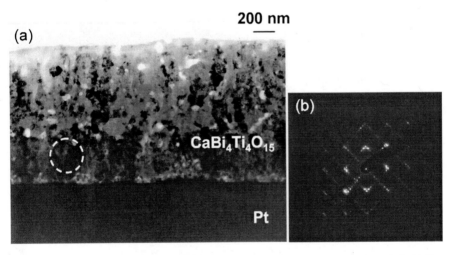

Figure 7   Cross-section TEM (a) and electron diffraction (b) profiles of 1 μm-thick CBTi144 film.

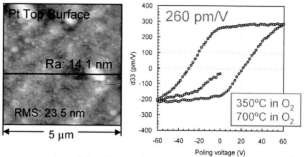

Figure 8   Surface topography of the Pt top electrode for the PFM measurements and d_{33}-V characteristic of 1 μm-thick CBTi144 film on Pt foil.

CONCLUSION

Polar-axis oriented CBTi144 films were fabricated on Pt foils using a complex metal alkoxide solution. Ambient during heat treatments such as pre-baking and crystallization affected the nucleation behaviors of the alkoxy-derived CBTi144 films. The pre-baking ambient affected the microstructure, and the crystallization ambient impacted the crystal perfection and the ferroelectric properties. The 500 nm-thick film that was pre-baked in air and then crystallized in oxygen flow consisted of columnar grains with rather rough surface and showed good ferroelectric properties. It was necessary for improvement of the electrical properties in thick films to obtain closely-packed dense structure with relatively smooth surface. The 1 μm-thick CBTi144 films through both pre-baking and crystallization in oxygen flow, maintained high polar-axis orientation and exhibited low leakage current density of 10^{-7} A/cm^2 and high piezoelectric constant, d_{33} of 260 pm/V.

REFERENCES

[1]E. C. Subbarao, J. Am. Ceram. Soc., **45**, 166 (1962).

[2]L. Korzanova, Ferroelectrics, **134**, 175 (1992).

[3]K. Kato, D. Fu, K. Suzuki, K. Tanaka, K. Nishizawa, T. Miki, Appl. Phys. Lett., **84**, 3771 (2004).

[4]K. Kato, K. Suzuki, K. Nishizawa and T. Miki, Appl. Phys. Lett. **78**, 1119 (2001).

[5]K. Kato, K. Suzuki, D. Fu, K. Nishizawa and T. Miki, Appl. Phys. Lett. **81**, 3227 (2002).

[6]D. Fu, K. Suzuki and K. Kato, Jpn. J. Appl. Phys., **42**, 5994 (2003).

[7]D. Fu, K. Suzuki and K. Kato, Appl. Phys. Lett., **85**, 3519 (2004).

[8]K. Kato, K. Tanaka, K. Suzuki, T. Kimua, K. Nishizawa, T. Miki, Appl. Phys. Lett., **86**, 112901 (2005).

[9]K. Kato, K. Tanaka, K. Suzuki, T. Kimua, K. Nishizawa, T. Miki, Integr. Ferroelectr., **80**, 21 (2006).

[10]K. Kato, K. Tanaka, Y. Guo, K. Suzuki, T. Kimua, K. Nishizawa, T. Miki, Integr. Ferroelectr., submitted.

[11]K. Kato, K. Suzuki, D. Fu, K. Nishizawa and T. Miki, Jpn. J. Appl. Phys., **41**, 6829 (2003).

[12]A. Z. Simoes, M. A. Ramirez, A. Ries, J. A. Varela, E. Longo, R. Ramesh, Appl. Phys. Lett., **88**, 072916 (2006).

[13]J. S. Lee, H. H. Kim, H. J. Kwon and Y. W. Jeong, Appl. Phys. Lett., **73**, 166 (1998).

PROCESSING OF POROUS $Li_{0.06}(Na_{0.5}K_{0.5})_{0.94}NbO_3$ CERAMICS AND THEIR PIEZOELECTRIC COMPOSITES WITH HETERO-CRYSTALS

Tomoya Imura, Ken-ichi Kakimoto
Department of Materials Science and Engineering, Graduate School of Engineering,
Nagoya Institute of Technology
Gokiso-cho, Showa-ku, Nagoya 466-8555, Japan

Katsuya Yamagiwa, Takeshi Mitsuoka, and Kazushige Ohbayashi
R&D Center, NGK Spark Plug Co., Ltd.
2808 Iwasaki, Komaki, Aichi 485-8510, Japan

ABSTRACT

$Li_{0.06}(Na_{0.5}K_{0.5})_{0.94}NbO_3$ (LNKN) pre-ceramic powder and phenol resin (Kynol^{TM}) fiber was mixed for fabrication of porous LNKN ceramics with different pore volumes. The porous LNKN ceramics were then converted to LNKN/KNbO$_3$ composites through soaking and heat-treatment by using a sol-gel precursor source of KNbO$_3$ composition. Dielectric and piezoelectric properties of the porous LNKN ceramics and LNKN/KNbO$_3$ composites were characterized and compared. The piezoelectric voltage coefficient (g_{33}) increased with the volume of porosity for the porous LNKN ceramics. Furthermore, the enhanced g_{33} value was measured for LNKN/KNbO$_3$ composite, and showed the maximum value of 63.0×10^{-3} Vm/N.

INTRODUCTION

Porous $Pb(Zr,Ti)O_3$ (PZT) ceramics and their polymer composites have been developed due to the possibility of obtaining high piezoelectric voltage coefficient (g_{33}) and effective acoustic impedance against dense monolithic ceramics.[1] At present, however, the toxicity of lead oxide lead to a demand for alternative lead-free piezoelectric materials, and the leading candidate is now focused on alkali-niobate systems; e.g., KNbO$_3$ and its relative $(Na_{0.5}K_{0.5})NbO_3$ systems.[2-5] Among them, Li-doped $(Na_{0.5}K_{0.5})NbO_3$ (LNKN) ceramics[6,7] have been considered as one of the best candidates for piezoelectric transducer application, since they show excellent piezoelectric properties such as piezoelectric constant d_{33}=200-235 pC/N, electromechanical planar and thickness coupling coefficients of k_p=38-44% and k_t=44-48%, and high Curie temperatures > 400°C for Li content of 5-7 mol%.[8]

In this study, porous $Li_{0.06}(Na_{0.5}K_{0.5})_{0.94}NbO_3$ (LNKN, 6 mol%Li) ceramics were fabricated by mixing of phenol resin fibers to the pre-ceramic powder, and sintering of the die-pressed samples. The derived porous LNKN ceramics was further converted to the composites containing KNbO$_3$ hetero-crystals inside the pores by using sol-gel precursor solution. The objective of this study is to characterize the structure and piezoelectric property of new composite ceramics of LNKN/KNbO$_3$ as well as porous LNKN ceramics.

EXPERIMENTAL

Preparation of porous ceramics

High-purity Li_2CO_3, Na_2CO_3, K_2CO_3 and Nb_2O_5 powders were weighed according to the compositional formula of $Li_{0.06}(Na_{0.5}K_{0.5})_{0.94}NbO_3$, and were wet ball-milled for 24 h using ethanol as a medium. The dried mixture was calcined at 850°C for 10 h. Phenol resin fibers

(Kynol^{TM}) were then mixed to the calcined powder at different content of 0, 10, 20 and 30 vol% as a pore-former. The disk-shaped samples with 12 mm in diameter were heated on an Al$_2$O$_3$ boat at 800°C for 2h to burn out the pore-former, and then cooled to the room temperature, Subsequently, the cooled specimens were placed on a powder-bed made of the same composition powders and then covered with an Al$_2$O$_3$ crucible. Finally, they were heated at the sintering temperature of 1082°C for 2 h.

Preparation of composites

The porous LNKN ceramics were soaked in a sol-gel precursor source of KNbO$_3$ composition with the concentration of 1.0 mol/l to 2-methoxyetanol. The soaked samples were dried and calcined at 150 and 450°C, respectively. These soaking, drying and calcining processes were performed 10 times repeatedly. The final samples were heated at 800°C for 1 h to form KNbO$_3$ crystals inside the pore.

Characterization

The microstructure was examined by scanning electron microscopy (SEM). Temperature dependence of dielectric constant was evaluated by using a LCR meter at temperature ranging from the room temperature to 550°C. The piezoelectric d$_{33}$ constant for the electrically poled specimens were measured by using a d$_{33}$ meter. The piezoelectric voltage coefficient (g$_{33}$) was calculated using the following formula:

$$g_{33} = d_{33} / \varepsilon^T_{33} \tag{1}$$

where ε^T_{33} is the dielectric constant measured parallel to the direction of the applied voltage for the electrically poled samples.

RESULTS AND DISCUSSION

Porous LNKN ceramics

Figure 1 shows the SEM images of the samples with different Kynol^{TM} fiber contents. It can be seen that cylindrical and anisotropic-shaped pores oriented randomly are formed in the ceramics. With increasing the content of the added Kynol^{TM} fiber, the porosity of the ceramics increases from 6 to 26%, as shown in Figure 2. It is noteworthy that the open porosity of the ceramics increases with Kynol^{TM} fiber content, but the closed porosity decreases in turn, compared with dense LNKN ceramics. This indicates that pore-bridging was much enhanced by using an anisotropic fiber shape of Kynol^{TM} fiber. In other words, it is considered that the porous ceramics prepared by using Kynol^{TM} fiber is suitable for impregnation of a sol-gel precursor solution into the pores.

The temperature dependence of the dielectric constant measured at 1 kHz for porous LNKN ceramics is shown in Figure 3. A peak observed at 448°C corresponds to the phase transition between cubic to tetragonal perovskite for LNKN ceramics, and this temperature was found to be independent with porosity. On the other hand, the dielectric constant measured at room temperature was decreased by the formation of pores (see in Figure 5).

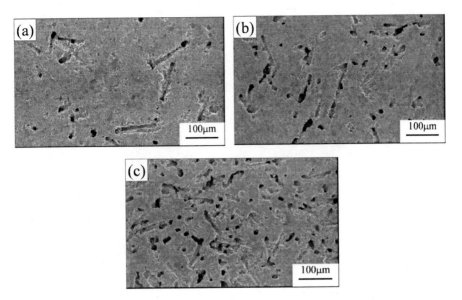

Figure 1. SEM images of the porous LNKN ceramics prepared by using
(a) 10, (b) 20 and (c) 30 vol% of Kynol^{TM} fiber.

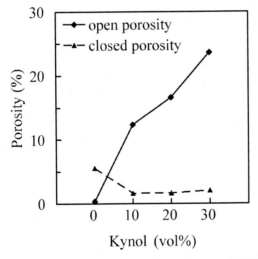

Figure 2. Variation of open and closed porosity in the porous LNKN ceramics
as a function of Kynol^{TM} fiber content.

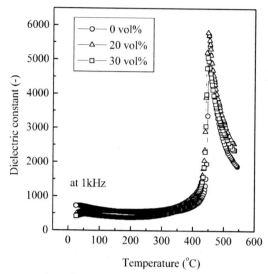

Figure 3. Temperature dependence of dielectric constant of the porous LNKN ceramics prepared by using (a) 0, (b) 20 and (c) 30 vol% of Kynol^{TM} fiber.

LNKN/KNbO$_3$ composite

Figure 4 presents the microstructure of typical surface and cross section of LNKN/KNbO$_3$ composite prepared from the porous LNKN ceramics using 30 vol% of Kynol^{TM} fiber. The surface exhibited uniform microstructure with KNbO$_3$ crystallized from its precursor gel, and almost no pores, compared with the microstructure shown in Figure 1(c). In contrast, the anisotropic-shaped pores, represented by single arrows in the figure, remained inside the composite. A part of them may be closed pores included in the starting material, but it can be confirmed that KNbO$_3$ crystals were also formed inside the pores, represented by double arrows in the figure.

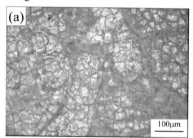 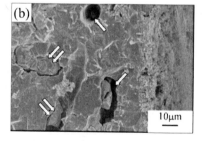

Figure 4. Microstructure of LNKN/KNbO$_3$ composite fabricated from the porous LNKN ceramics (source: 30 vol% Kynol^{TM} fiber): (a) surface , (b) cross section.

The variation of piezoelectric properties of the porous LNKN ceramics and LNKN/KNbO$_3$ composites is shown in Figure 5. The dielectric constant of the porous LNKN ceramics decreases sharply from 730 to 392 with increasing Kynol^{TM} fiber content from 0 to 10 vol%, whereas no significant change is observed in its ranging from 10 to 30 vol%. This may seem to be correlated with the mixture effect of microstructure, pore-size distribution and interconnectivity (open/closed porosity).[9-11] With respect to dielectric constant, there is not so much difference between the porous LNKN ceramics and LNKN/KNbO$_3$ composites.

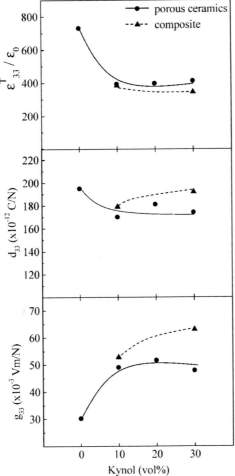

Figure 5. Variation of piezoelectric properties as a function of Kynol^{TM} fiber content for porous LNKN ceramics and LNKN/KNbO$_3$ composites.

On the other hand, the d_{33} constant of porous LNKN ceramics decreases moderately with increasing Kynol™ content. According to the variations of dielectric constant and d_{33} constant, the calculated g_{33} constant of porous LNKN ceramics increases with increasing Kynol™ content. On the other hand, the value of d_{33} for the LNKN/KNbO$_3$ composite (30vol% Kynol™) shows 192×10^{-12} C/N against 174×10^{-12} C/N for the porous LNKN ceramics, and the derived g_{33} values for this composite is 63.0×10^{-3} Vm/N. The enhanced d_{33} constant observed in the composite seems to be due to the effect of the formation of KNbO$_3$ crystals filled inside the pores, but the detailed mechanism for the enhanced property will be reported in the future.

SUMMARY

The porous Li$_{0.06}$(Na$_{0.5}$K$_{0.5}$)$_{0.94}$NbO$_3$ (LNKN) ceramics and LNKN/KNbO$_3$ composites were prepared, and their dielectric and piezoelectric properties were compared. The piezoelectric voltage coefficient (g_{33}) of the porous LNKN ceramics increased with the volume of porosity. The g_{33} value of LNKN/KNbO$_3$ composite showed the maximum value of 63.0×10^{-3} Vm/N.

REFERENCES

[1]N. M. Gokhale, S. C. Sharma, and R. Lal, "Fabrication of PZT-Polymer Composite Materials Having 3-3 Conectivity for Hydrophone Applications," *Bull. Mater. Sci.*, **11**, 49-54 (1988).

[2]G. Shirane, R. Newnham, and R. Pepinsky, "Dielectric Properties and Phase Transitions of NaNbO$_3$ and (Na,K)NbO$_3$," *Phys. Rev.*, **96**, 581-88 (1954).

[3]L. Egerton, and D. M. Dillon, "Piezoelectric and Dielectric Properties of Ceramics in the System of Potassium-Sodium Niobate," *J. Am. Ceram. Soc.*, **42**, 438- (1959).

[4]R. E. Jaeger, and L. Egerton, "Hot Pressing of Potassium Sodium Niobates," *J. Am. Ceram. Soc.*, **45**, 209-13 (1962).

[5]G. H. Haertling, "Properties of Hot-Pressed Ferroelectric Alkali Niobate Ceramics," *J. Am. Ceram. Soc.*, **50**, 329-30 (1967).

[6]E. Hollenstein, M. Davis, D. Damjanovic, and N. Setter, "Piezoelectric Properties of Li- and Ta-Modified (K$_{0.5}$Na$_{0.5}$)NbO$_3$ Ceramics," *Appl. Phys. Lett.*, **87**, 182905 (2005).

[7]T. Fu-sheng, D. Hong-liang, L. Zhi-min *et all.*, "Preparation and Properties of (K$_{0.5}$Na$_{0.5}$)NbO$_3$-LiNbO$_3$ Ceramics," *Trans. Nonferrous Met. Soc. China*, **16**, s466-69 (2006).

[8]Y. Guo, K. Kakimoto, and H. Ohsato, "Phase Transitional Behavior and Piezoelectric Properties of (Na$_{0.5}$K$_{0.5}$)NbO$_3$-LiNbO$_3$ Ceramics," *Appl. Phys. Lett.*, **85**, 4121-23 (2004).

[9]E. Roncari, C. Galassi, F. Cracium, C. Capiani, and A. Piancastelli, "A Microstructural Study of Porous Piezoelectric Ceramics Obtained by Different Methods," *J. Eur. Ceram. Soc.*, **21**, 409-17 (2001).

[10]Z. He, J. Ma, and R. Zhang, "Investigation on the Microstructure and Ferroelectric Properties of Porous PZT Ceramics," *Ceram. International*, **30**, 1353-56 (2004).

[11]T. Hayashi, S. Sugihara, and K. Okazaki, "Processing of Porous 3-3 PZT Ceramics Using Capsule-Free O$_2$-HIP," *Jpn. J. Appl. Phys.*, **30**, 2243-46 (1991).

ELECTRIC-FIELD-INDUCED DIELECTRIC, DOMAIN AND OPTICAL PHENOMENA IN HIGH-STRAIN $Pb(In_{1/2}Nb_{1/2})_{1-x}Ti_xO_3$ (x=0.30) SINGLE CRYSTAL

C.-S. Tu[1], C.-M. Hung[1], R.R. Chien[2], V.H. Schmidt[2], and F.-T. Wang[1]
[1]Graduate Institute of Applied Science and Engineering, Fu Jen Catholic Univ., Taipei 242, ROC
[2]Department of Physics, Montana State University, Bozeman, Montana 59717, USA

ABSTRACT

Dielectric permittivity, domain structure, and electric polarization have been measured as a function of temperature in a (001)-cut $Pb(In_{1/2}Nb_{1/2})_{0.7}Ti_{0.3}O_3$ (PINT30%) crystal before and after an electric-field poling. After a prior poling long-range percolating domains were induced with disappearance of frequency-dependent dielectric relaxation. As temperature increases, the long-range percolations were broken near 390 K where an extra dielectric anomaly was observed. The dielectric relaxation behavior reappears entirely near 415 K upon heating. The dielectric permittivity was found to follow the Curie-Weiss behavior above the Burns temperature. Thermodynamics of field-induced percolation plays an important role in phase thermal instability.

INTRODUCTION

High-strain ferroelectric $Pb(Mg_{1/3}Nb_{2/3})_{1-x}Ti_xO_3$ (PMNT) and $Pb(Zn_{1/3}Nb_{2/3})_{1-x}Ti_xO_3$ (PZNT) crystals have demonstrated their value in piezoelectric and medical ultrasonic imaging devices because of high piezoelectric response.[1-3] However, thermal instability (variation with temperature) caused by overheating remains a vital issue in use of these materials. Physical properties of PMNT and PZNT are sensitive to Ti content, poling process, electric (E)-field strength, crystallographic orientation, and history.[4,5] It has been a goal to find high-strain piezoelectric crystals with high Curie temperature (T_C) and to avoid thermal instability. Among high-T_C piezoelectric crystals, $Pb(In_{1/2}Nb_{1/2})_{1-x}Ti_xO_3$ (PINTx) has drawn attention in recent years.[6-12]

From temperature-, E-field-, and pressure-dependent dielectric results, phase diagrams of E field and pressure versus temperature were proposed for PINT single crystals near the morphotropic phase boundary.[6] The (001)-cut PINT34% single crystal (starting composition) grown by the modified Bridgman method with a PMNT29% seed crystal, exhibits higher piezoelectric constant ($d_{33} > 2000$ pC/N) and electromechanical coupling factor ($k_{33} \cong 94\%$) than those obtained from the (011)- and (111)-cut crystals.[8] The (001)-cut PINT35% (starting composition) crystal grown with a PMNT33% seed crystal, has $d_{33}=2000$ pC/N and $T_C=269$ °C.[9] A diffuse rhombohedral (R)→tetragonal (T) transition was seen in a (001)-cut PINT28% crystal under a dc E field along [010] at room temperature (RT).[10] R and T phase domains were seen to coexist in the range of 6-12 kV/cm.[10] A diffuse frequency-dependent transition was observed in a (001)-cut PINT30% single crystal associated with polar nanoclusters below the Burns temperature.[11] The unpoled (001)-cut PINT30% crystal shows no birefringence, indicating that the average structural symmetry is optically isotropic.[11] A first-order-type tetragonal (T)/monoclinic (M)→cubic (C) transition was observed in an unpoled (001)-cut PINT40% at $T_C \cong 486$ K below which T and M domains coexist.[12] A coexistence of T and possible M phases was proposed in PINT37% ceramics.[13]

EXPERIMENTAL METHODS

In this study, the single crystal PINT30% (starting composition) was grown using a modified Bridgman method, in which an allomeric PMNT29% seed crystal was used.[7] The allomeric seed crystal can restrain the formation of pyrochlore phase and improves spontaneous nucleation.[7] By X-ray fluorescence analysis, MgO was found in the crystal due to ionic migration. For dielectric measurements, a Wayne-Kerr Analyzer PMA3260A was used to obtain the real part ε' of dielectric permittivity. The sample was cut perpendicular to the <001> direction and its dimensions are $2.3 \times 2.2 \times 0.5$ mm^3. Gold electrodes were deposited on sample surfaces by dc sputtering. Two processes were mainly used in the dielectric study. The first is called "zero-field-heated" (ZFH), in which the data were taken upon heating without any poling. In the second process "prior-poled before zero-field-heated" (PP-ZFH), the sample was poled at RT with a dc E field of 5 kV/cm along [001] for 1 hr, then cooled to 140 K without an E field before ZFH was performed. An irregular piezoelectric resonance was observed for frequency $f > 100$ kHz in the PP-ZFH dielectric spectra. Hysteresis loops were taken by using a Sawyer-Tower circuit at f=46 Hz. A Janis CCS-450 cold-head was used with a Lakeshore 340 controller for temperature-dependent measurements.

RESULTS AND DISCUSSION

Domain structures were observed by using a Nikon E600POL polarizing microscope with a crossed polarizer/analyzer (P/A) pair. Transparent conductive films of indium tin oxide (ITO) were deposited on the (001) basal surfaces by dc sputtering. The sample thickness is about 70 μm. Angles of the P/A pair measured in this work are with regard to the [110] direction. The experimental configuration of E-field dependent domain observation can be found in Ref. 4.

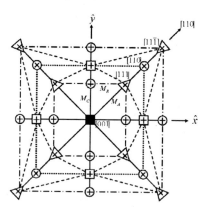

Figure 1 Relation between the optical extinction orientations corresponding to polarizations of various domains, projected on the (001) plane.

Figure 1 shows the (001)-cut projection (with all four sides folded out) of relations among the various phases (or symmetries) and corresponding polarizations (**P**). Squares, triangles, and circles indicate the directions of T, R, and orthorhombic (O) polarizations, respectively. Solid crossed lines inside the symbols indicate that domains will have extinction for optical E field along the radial and circumferential axes. Solid, dash-dot, dashed and dotted lines indicate directions that polarizations can take for various M cells. The M_C cell **P** lies between two adjacent T and O **P** vectors. The M_A cell has **P** between two adjacent T and R **P** vectors, whereas the M_B cell has **P** between two adjacent R and O **P** vectors. The central "black" square indicates total optical extinction. More details for using optical extinction to determine domain phases can be found in Ref. 4.

Figure 2 shows dielectric results obtained from PP-ZFH and ZFH (inset) processes for $f \leq 100$ kHz. Frequency-dependent maxima were observed in the region of ~430-460 K and their peaks shift toward higher temperatures with increasing frequency. In the PP-ZFH an extra dielectric anomaly (that was not seen in the ZFH) occurs near 390 K, below which the frequency-dependent relaxation behavior was diminished. The disappearance of relaxation behavior below 390 K in PP-ZFH implies formation of long-range percolation clusters and reduction of domain walls due to E-field poling. An extra peak was also observed in $Pb(Mg_{1/3}Nb_{2/3})O_3$ (PMN) in the field-heating dielectric result and was connected to the percolating clusters due to the suppression of the random fields.[14]

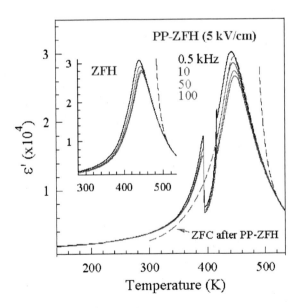

Figure 2 ZFH and PP-ZFH dielectric permittivity. The red dashed curve indicates the ZFC dielectric data after PP-ZFH at f=10 kHz.

As temperature increases in PP-ZFH, the frequency-dependent dielectric behavior reappears entirely near 415 K (Fig. 2). Above T_m which corresponds to the dielectric maximum, ε' roughly follows Curie-Weiss behavior. The blue dashed lines are fittings to the Curie-Weiss equation, $\varepsilon'(T)=C/(T-T_o)$, with $C=3.9\times10^5$ and $T_o=472$ K for ZFH, and $C=3.9\times10^5$ and $T_o=477$ K for PP-ZFH. Noticeable deviations from the Curie-Weiss law begin near 510 and 520 K for ZFH and PP-ZFH, respectively. We consider 510 and 520 K to be the Burns temperatures (T_B), below which polar nanoclusters start to develop. Their weaker dielectric response causes deviations from the Curie-Weiss law and their dynamics are responsible for the broad dielectric relaxation behavior.[15] As shown by the red dashed curve in Fig. 2, the zero-field-cooled (ZFC) dielectric permittivity measured after the PP-ZFH does not show the "extra" anomaly. This indicates that the poling effect is not thermally stable and can be erased easily by thermal annealing.

Figure 3 shows (a) temperature-dependent hysteresis loops of polarization versus E field and (b) remanent polarization (P_r) and coercive field (E_C) versus temperature, respectively. The spontaneous polarization (P_S), P_r, and E_C taken at RT are ~22 μC/cm², ~20 μC/cm², and ~3.4 kV/cm, respectively. As shown in Fig. 3(b), P_r and E_C exhibit a clear dip anomaly near 380 K, implying a significant decline of electric polarization. Note that 380 K is fairly consistent with the temperature where the extra dielectric anomaly occurs in the PP-ZFH.

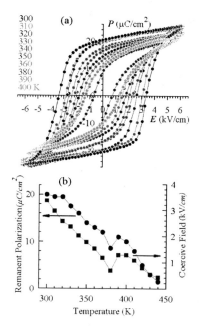

Figure 3 Temperature-dependent (a) hysteresis loops and (b) remanent polarization and coercive field.

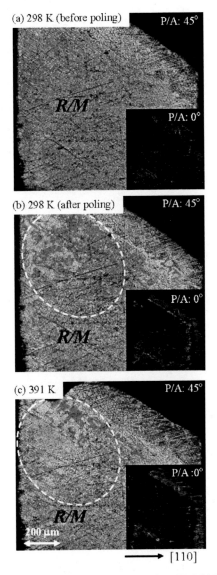

Figure 4 Domain observations before and after a prior poling (E=5 kV/cm) at RT. Domain structures were viewed at P/A:45° and 0° (insets).

Before a prior poling, the sample mostly shows an optical extinction at P/A:0° at 298 K as shown in Fig. 4(a), indicating possible R and M (M_A- or M_B-type) domains according to Fig. 1. This coexistence of R and M domains is confirmed by the <002> X-ray diffraction which shows a sharp peak (R phase) and a broad shoulder (M phase) as seen in Fig. 5. A similar X-ray profile was observed in PMNT31% ceramic which has a coexistence of R and M domains at RT.[3] ZFH domain structures after poling at E=5 kV/cm along [001] are illustrated in Figs. 4(b) and 4(c) for 298 and 391 K. After poling at 298 K, long-range domains (or clusters) were induced with extinction angle at P/A:0° as encircled in Fig. 4(b). The <002> X-ray diffraction (Fig. 5) shows that the ratios of the intensities of the shoulder and the sharp peak are almost the same before and after poling at E=5 kV/cm, which means that amount of the M domains does not change by E-field poling. As temperature increases, birefringences of the E-field-induced long-range domains become weaker radically near 391 K [Fig. 4 (c)] with the same extinction at P/A:0°. No obvious domain anomaly was observed around 415 K where the frequency-dependent dielectric behavior reenters the crystal entirely as seen in Fig. 2. Note that the maximal resolution of the polarizing microscope is about one micrometer (μm) due to optical diffraction limit. These phenomena suggest that thermal energy triggers breakdown of long-range percolations near 390 K, where nanoclusters begin to reestablish and reduce the average anisotropic birefringence. Above 415 K, nanoclusters integrate into the crystal and meanwhile their dynamics reestablish the frequency-dependent dielectric relaxation. The domain matrix displays total optical extinction near 453 K, indicating a cubic (C) phase.

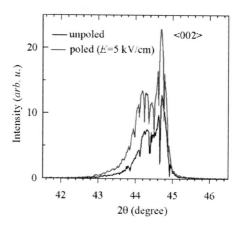

Figure 5 <002> X-ray diffraction profiles before and after a prior poling (E=5 kV/cm).

The (001)-oriented E-field-dependent domain observation at RT shows that only a partial domain matrix could be poled to the [001] tetragonal phase and the [001] monodomain cannot be achieved at poling field E=40 kV/cm along [001]. The [001] monodomain also could not be reached in a (001)-cut PMNT24% crystal under E=44 kV/cm along [001].[4]

CONCLUSIONS

After a prior E-field poling along [001], long-range percolating domains (or polar clusters) were induced macroscopically with disappearance of the frequency-dependent dielectric relaxation behavior. As temperature increases, the long-range percolations were broken down near 390 K where polar nanoclusters begin to reenter the system. The percolation breakdown is activated by thermal energy and is responsible for the extra dielectric anomalies which were not seen in the unpoled crystal. Polar nanoclusters completely integrate into the crystal near 415 K and meanwhile the dielectric relaxation reappears entirely as seen in the ZFH. The crystal enters into the cubic phase near 453 K with existence of polar nanoclusters which vanish above $T \geq T_B \cong 520$ K. A similar E-field-induced extra dielectric anomaly was seen in PMNT24% crystal and was connected with cluster percolations.[16] It seems that thermodynamics of E-field-induced percolation plays a vital role in temperature-dependent phase transitions of relaxor-based ferroelectrics.

The authors would like to thank Dr. H. Luo for crystals. This work was supported by National Science Council of Taiwan Grant No. 95-2112-M-030-001.

REFERENCES

[1] J. Kuwata, K. Uchino, and S. Nomura, "Phase transitions in the $Pb(Zn_{1/3}Nb_{2/3})O_3$-$PbTiO_3$ system," *Ferroelectrics* **37**, 579-582 (1981).

[2] T.R. Shrout, Z.P. Chang, N. Kim, and S. Markgraf, "Dielectric behavior of single crystals near the $(1-x)Pb(Mg_{1/3}Nb_{2/3})O_3$-$(x)PbTiO_3$ morphotropic phase boundary," *Ferroelectrics Letters Sect.* **12**, 63-69 (1990).

[3] B. Noheda, D.E. Cox, G. Shirane, J. Gao, and Z.-G. Ye, "Phase diagram of the ferroelectric relaxor $(1-x)Pb(Mg_{1/3}Nb_{2/3})O_3$-$xPbTiO_3$," *Phys. Rev.* B **66**, 054104/1-10 (2002).

[4] R.R. Chien, V.H. Schmidt, C.-S. Tu, L.-W. Hung, and H. Luo, "Field-induced polarization rotation in (001)-cut $Pb(Mg_{1/3}Nb_{2/3})_{0.76}Ti_{0.24}O_3$," *Phys. Rev.* B **69**, 172101/1-4 (2004).

[5] C.-S. Tu, R.R. Chien, F.-T. Wang, V.H. Schmidt, and P. Han, "Phase stability after an electric-field poling in $Pb(Mg_{1/3}Nb_{2/3})_{1-x}Ti_xO_3$ crystals," *Phys. Rev.* B **70**, 220103/1-4 (2004).

[6] N. Yasuda, H. Ohwa, K. Ito, M. Iwata, and Y. Ishibashi, "Dielectric properties of the $Pb(In_{1/2}Nb_{1/2})O_3$-$PbTiO_3$ single crystal," *Ferroelectrics* **230**, 115-120 (1999).

[7] Y. Guo, H. Luo, T. He, X. Pan, and Z. Yin, "Peculiar properties of a high curie temperature $Pb(In_{1/2}Nb_{1/2})O_3$-$PbTiO_3$ single crystal grown by the modified Bridgman technique," *Solid State Commun.* **123**, 417-420 (2002).

[8] Y. Guo, H. Luo, T. He, X. Pan, and Z. Yin, "Electric-field-induced strain and piezoelectric properties of a high Curie temperature $Pb(In_{1/2}Nb_{1/2})O_3$-$PbTiO_3$ single crystal," *Materials Research Bulletin* **38**, 857-864 (2003).

[9] Z. Duan, G. Xu, X. Wang, D. Yang, X. Pan, and P. Wang, "Electrical properties of high Curie temperature $(1-x)Pb(In_{1/2}Nb_{1/2})O_3$-$xPbTiO_3$ single crystals grown by the solution Bridgman technique," *Solid State Commun.* **134**, 559-603 (2005).

[10] N. Yasuda, M. Sakaguchi, Y. Itoh, H. Ohwa, Y. Yamashita, M. Iwata, and Y. Ishibashi, "Effect of electric fields on domain structure and dielectric properties of $Pb(In_{1/2}Nb_{1/2})O_3$-$PbTiO_3$ near morphotropic phase boundary," *Jpn. J. Appl. Phys.* **42**, 6205-6208 (2003).

[11]C.-S. Tu, C.-M. Hung, F.-T. Wang, R. R. Chien, and S.-W. Yang, "Dielectric and optical behaviors in relaxor ferroelectric $Pb(In_{1/2}Nb_{1/2})_{1-x}Ti_xO_3$ crystal," *Solid State Commun.* **138** (4), 190-193 (2006).

[12]C.-S. Tu, F.T. Wang, C.M. Hung, R.R. Chien, and H. Luo, "Dielectric, domain, and optical studies in high-Curie-temperature $Pb(In_{1/2}Nb_{1/2})_{1-x}Ti_xO_3$ (x=0.40) single crystal," *J Appl. Phys.* **100** 104104/1-5 (2006).

[13]C. Augier, M. PhamThi, H. Dammak, and P. Gaucher, "Phase diagram of high T_c $Pb(In_{1/2}Nb_{1/2})O_3$–$PbTiO_3$ ceramics," *J. Europ. Ceramic Soc.* **25**, 2429-2432 (2005).

[14]V. Westphal, W. Kleemann, and M. D. Glinchuk, "Diffuse phase transitions and random-field-induced domain states of the relaxor ferroelectric $PbMg_{1/3}Nb_{2/3}O_3$," *Phys. Rev. Lett.* **68**, 847-850 (1992).

[15]D. Viehland, S.J. Jang, L.E. Cross, and M. Wuttig, "Deviation from Curie-Weiss behavior in relaxor ferroelectrics," *Phys. Rev.* B **46**, 8003-8006 (1992).

[16]C.-S. Tu, C.L. Tsai, V.H. Schmidt, H. Luo, and Z. Yin, "Dielectric, hypersonic and domain anomalies of $(PbMg_{1/3}Nb_{2/3}O_3)_{1-x}(PbTiO_3)_x$ single crystals," *J. Appl. Phys.* **89**, 7908-7916 (2001).

EFFECTS OF ELECTRIC FIELD ON THE BIAXIAL STRENGTH OF POLED PZT

Hong Wang and Andrew A. Wereszczak
Ceramic Science and Technology
Materials Science and Technology Division
Oak Ridge National Laboratory
Oak Ridge, TN 37831

ABSTRACT

The mechanical integrity of piezoelectric ceramics plays a crucial role in the performance and design of lead zirconate titanate (PZT) piezo stack actuators especially as PZT actuators become physically larger and are sought to operate under harsher conditions. Reliable design of such systems demands additional consideration of a number of issues including electro-mechanical coupling, as well as strength-size scaling. This study addresses some of those issues through the use of ball-on-ring (BoR) equibiaxial flexure strength tests of two PZT piezo ceramics, subjected to different electric fields. Fracture surfaces and failure origins were analyzed using optical and scanning electronic microscopy. The both sign of the electric field and its magnitude can alter the two-parameter Weibull distribution. These results serve as input data for future probabilistic reliability analysis of multilayer PZT piezo actuators.

INTRODUCTION

The implementation of a lead zirconate titanate (PZT) piezo fuel injection system can reduce pollutants such as NOx and particulate matter (PM) by 30%, fuel consumption by 15% and engine noise by 6 dB. However, it is recognized that the performance of multilayer PZT actuators can be severely degraded by electromechanical fatigue, so making their successful utilization successful is not a trivial endeavor. The failure and reliability of piezo actuators has been a subject of study since the late 1980s [1, 2, 3, 4]. The fracture of the PZT layer has been identified as one of the prime sources of failure of multilayer actuators, along with delamination and arcing. Though PZT ceramics have undergone continued maturation and improvement since then, there is still an insufficient source of mechanical data available for probabilistic design of PZT actuators. This paper will describe our experimental work that addresses some of the above-mentioned issues. In the following, a brief introduction of the experimental set-up and materials tested are given followed by the strength result including Weibull analysis and fractography.

EXPERIMENTAL APPROACH

Materials Tested and Specimen Preparation

Two commercially available materials were evaluated in this study - PSI 5A4E and PSI 5H4E (Piezo systems, Inc., Cambridge, MA). Relevant parameters of the two materials are listed in Table 1. It can be seen that 5A4E has a higher coercive field (E_c) than 5H4E, while its piezoelectric coefficient (d_{33}) is lower than 5H4E. Noticeably, 5A4E has a high Curie temperature ($T_c = 350^\circ C$), which is considered to be an excellent candidate for diesel fuel injector systems.

As-received sheets of the two PZT materials having dimensions of 72.39 x 72.39 x 0.267 mm were Ni-electroded and fully poled by the manufacturer. These sheets were cut into small size plates (Bomas Specialty Machining, Somerville, MA) for the ball-on-ring (BoR) tests. For each sheet, there were a total of 49 square specimens obtained with a nominal size of 10.00 x 10.00 mm. The surface roughness (Ra) value of the as-received plates was ~ 0.8 - 0.9 mm as measured with a laser profilometer.

Table 1. PZT materials PSI 5A4E and PSI 5H4E were studied.

Piezo Systems			PSI-5A4E*	PSI-5H4E*
Industry designations			Navy II, 5A	Navy VI, 5H
Dielectric const	$\varepsilon_{33}^T / \varepsilon_0$ **		1,800	3,800
Piezoelectric coeff.	d_{31}	pC/N	-190	-320
	d_{33}	pC/N	390	650
Coercive field	E_C	kV/mm	1.2	0.8
Elastic modulus	Y_{11}^E	GPa	66	62
	Y_{33}^E	GPa	52	50
Curie temperature	T_C	°C	350	230

* Manufacturer reported data;
** ε_0 = the permittivity of vacuum or air, 8.854*10^{-12} F/m.

Experimental Set-up
    The ball-on-ring set-up is modified to include the ability to apply a range of high voltages as shown in Fig. 1. The bottom center piece is a steel supporting ring (sized with 7.444 mm ID, 25.400 mm OD and 6.35 mm thickness) with a high voltage cable attached. This support ring serves as an electrode and rests on (electrically insulating) alumina plate. On the top of the support ring, the 10.00 x 10.00 mm PZT specimen is placed. Above the specimen is a loading steel ball with 2.00 mm diameter and that also serves as the other electrode. The loading ball is mounted onto the bottom of a steel spindle using conductive resin (TIGA Silver 901, Resin Tech, S. Easton, MA). Above the spindle are an electrically insulating polymer adapter and a 1000g load cell. The (vertical) Z-direction loading action was controlled by a stepper motor. X and Y coordinates of the base are controlled by other two stepper motors. Therefore, the alignment of the loading axis with the supporting (ring) axis is controllable within several micrometers. The mounting process of the PZT specimen onto the support ring is also monitored with a TV screen through a video camera. The control and calibration of system and data acquisition are controlled using LabView software (NI, Austin, TX).

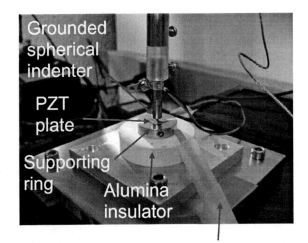

HV cable

Fig. 1 Experimental set-up details for superimposed mechanical and electric field testing. The HV (high voltage) cable and ground wire are connected into a HV amplifier that is not shown.

The high voltage (HV) cable and the ground wire are connected to the respective terminals of a high voltage amplifier (Trek model 609E-6, Medina, NY) that is powered by a DC power supply.

Test Procedures

A total of four combinations of electro-mechanical loading were tested in this study. The first two correspond to an open circuit (OC) field with the positive electrode attached to the tension side of the specimen (PoT) and with the negative electrode attached to the tensile surface (NoT), and the second pair correspond to a high electric field with the PoT and the NoT. The existing electrical field in the OC is typically low, and therefore, the first two combinations would exploit any surface condition effect on strength (if it exists), but not the electrical field effect. With a high electric field (E) applied, a positive E was actually generated in the PoT, and a negative E was generated in the NoT due to the configuration of the set-up. In each case of the latter two combinations, the level of E applied was equivalent to the respective coercive field E_c of each PZT material; namely, 1.2 kV/mm for 5A4E and 0.8 kV/mm for 5H4E. For each of four combinations, nine to fifteen specimens were tested to detect a statistically significant effect for a specific PZT. For the 5H4E, remnant specimens from a previous period of our project were also tested because it was found that the strength of one sheet exhibited substantial differences. The load was applied through displacement-controlled mode with a cross head speed of 0.001 mm/s.

The failure of the poled PZT plates was not catastrophic and the load-time curves exhibited multiple peaks. The some of tests were interrupted in order to inspect the damage sequences associated with those peaks. Both optical (Nikon Nomarski Measure Scope MM-11, Tokyo, Japan) and SEM (Hitachi S4100 field emission scanning electron microscope, San Jose, CA) were used to examine the recovered plate after the first peak. In most cases, the optical microscope was capable of catching the primary crack on the tensile side of the recovered plate. However, there were circumstances in which both optical microscope and SEM failed to reveal that cracking. Therefore, to study this further the Ni-electrode was removed using a chemical solution of 75% nitric acid + 25% water for 1.5 hours to expose cracks on the tension side of the specimen. As a result of this investigation, the load at the first peak was thereafter taken as the failure load of the material and used in the calculation of biaxial strength [5].

The strength data set for each loading condition was subsequently analyzed using Weibull statistics software (WeibPar, Connecticut Reserve Technologies, Inc., Strongsville, OH). The software estimated Weibull modulus (m) and characteristic strength (σ_θ) and their confidence intervals. Unbiased maximum likelihood parameter estimation was employed for these parameters as recommended by ASTM Standard C1239-00 [6]. Lastly, the fractographical studies were conducted according to ASTM Standard C1322-96a [7].

EXPERIMENTAL RESULTS

Weibull Parameters and Confidence Rings
Figure 2 shows the strength results of all the specimens cut from 5A4E Sheet 3. In the left column [Fig. 2(a)] are those with the OC condition where the upper plot is for samples tested with the PoT electrode configuration and the lower plot represents data for the NoT configuartion. For each electro-mechanical load case, the WeibPar output is also given in Fig. 2(a) along with the parameters used in the test, including characteristic strength (σ_θ) and Weibull modulus (m) and corresponding 95% confidence intervals. As seen, these two data sets are essentially overlaid, and therefore, the strength of this material was independent of which side was loaded in tension.

In the right column [Fig. 2(b)] is the specimen strength data from same sheet (Sheet 3) with concurrently applied electrical field. Under a positive E, the characteristic strength was found to be 158 MPa and that was statistically significantly higher than that of the OC. Under a negative E, the characteristic strength was 116 MPa and statistically lower than that of the OC. Consequently, a positive electric field enhances the strength and a negative one reduces it. The Weibull modulus in each of the cases was slightly lower; however, this was not significant.

As mentioned above, both the data sets of PoT and NoT under the OC were not different; therefore, the data were pooled. The pooled data set was then re-evaluated and the parameters re-estimated. Figure 3 illustrates the confidence rings of estimates (with Weibull modulus in abscissa and characteristic strength in ordinate) for three cases. The lower ring is for E = -1.2 kV/mm, the middle one is for the OC, and the upper one for E = 1.2 kV/mm. The trend shows the ring shifts downwards under a negative field, and upwards under a positive field. These shifts in strength distributions are statistically significant.

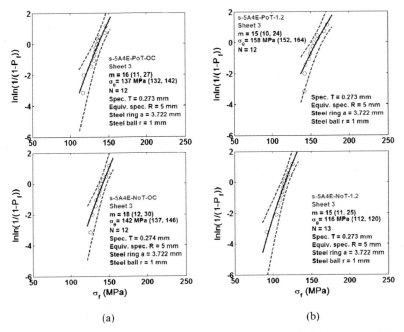

Fig. 2 Weibull analysis results for 5A4E; (a) gives the results for PoT (upper) and NoT (lower) under OC condition, and (b) for E = 1.2 kV/mm (upper) and -1.2 kV/mm (lower).

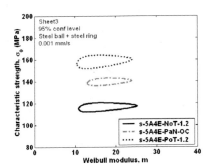

Fig. 3 Confidence rings for 5A4E. The confidence ring s-5A4E-PaN-OC represents the data set pooled for square plates of 5A4E with PoT and NoT under OC (Table 2).

For 5H4E, the related Weibull parameters are listed in Table 2 where the intervals correspond to 95% confidence levels unless indicated otherwise. The confidence ring now shifts downward and to the left for E = -0.8 kV/mm and upward and to the right when E = 0.8 kV/mm (Table 2). These shifts are statistically significant. This observation indicates that not only the characteristic strength changes, but also the Weibull modulus exhibits a similar tendency. Additionally, a significant sheet-to-sheet variation of characteristic strength was also observed (Table 2).

Overall, it seems difficult to see a definite trend with respect to the electrical field effect on the Weibull modulus where the E effect is negligible for 5A4E but strong for 5H4E. However, the data on the characteristic strength certainly displays a consistent trend for these PZT materials. For 5A4E, the strength decreased 17% under the negative E and increased 13% under the positive E. For 5H4E, the strength of Sheet 3 showed a decrease (18%) and an increase (21%) under the respective electric field, while that of Sheet 1 exhibited a more significant decrease (27%) under the negative field than its increase (12%) under the positive field.

Table 2 Weibull analysis results for two PZT materials– PZT 5A4E and PZT 5H4E

PZT (PSI)	Sheet	Tension side	E (kV/mm)	# of tests	Char. Strength (MPa)		Weibull modulus	
					σ_θ *	Interval	m*	Interval
5A4E	Sheet 3	PoT	OC	12	137	132, 142	16	11, 27
		NoT	OC	12	142	137, 146	18	12, 30
		Pooled	OC	24	140	136, 143	17	13, 24
		PoT	1.2	12	158	152, 164	15	10, 24
		NoT	1.2	13	116	112, 120	15	11, 25
5H4E	Sheet 1	PoT**	OC	6	135	131, 139	20	15, 41
		NoT	OC	7	141	135, 147	17	11, 34
		Pooled	OC	13	139	134, 143	18	13, 28
		PoT**	0.8	7	151	148, 155	27	18, 56
		NoT	0.8	7	102	91, 113	7	5, 14
	Sheet 3	PoT**	OC	15	99	95, 102	11	9, 17
		NoT	OC	9	106	97, 114	8	6, 15
		Pooled	OC	24	102	97, 106	9	7, 13
		PoT**	0.8	10	123	120, 127	17	12, 28
		NoT	0.8	15	84	78, 90	8	6, 12

*Point estimate was based on unbiased maximum likelihood estimation.
**Interval corresponds to 90% confidence level.

Fractographical Studies

Because strength exhibited a significant decrease due to the negative electrical field, the case of E = -E$_c$ became the focus of the fractographical study of both PZT materials. The image on the fracture surface in Fig. 4 (a) was taken for a 5A4E specimen subjected to E = -1.2 kV/mm, which had an intermediate strength. The bottom of the cross section was the tension side and the top was the compression side. An arch pattern shows up in the background that consists of many arch ledges and that laterally spreads. These ledges converge into the failure origin on the tension side. The size and shape of failure origin was delineated by the arrows. It is hard to classify the observed flaw according to ASTM Standard [7], but it could be appropriately designated as a surface-type processing flaw, or simply, surface processing flaw. The enlarged view on the failure origin, in Fig. 4 (b), revealed the prevailing transgranular fractures. In fact, it is the adjoining transgranular fractures that form the ledges in the background. On the other fracture surfaces (not shown), the cracked grained boundaries, volume pores and surface voids were observed within or around the failure origin. Overall, the EDS did not reveal any significant chemical difference between the failure origin and the background.

For 5H4E sheet 3, a prevailing intergranular fracture was revealed on the fracture surface. Because of that, the change of topographical features between the background and the failure origin on the tensile edge becomes quite subtle. The SEM image illustrated that the ledges on the fracture surface were not as distinct as those in 5A4E. However, a surface-type processing flaw is still identified as the type of failure origin.

DISCUSSIONS

The obtained biaxial (characteristic) strengths of PSI 5A4E and 5H4E are 137 to 142 MPa and 99 to 141 MPa under OC (Table 2), respectively. These data are generally higher than the values available in the literature. Jaffe and Berlincourt [8] reported a tensile strength of 75.8 MPa for both PZT-5A and -5H, but they did not give any experimental detail. In another study, Zhoga and Shipeizman [9] obtained a very low strength of 20 MPa on PZT-19 using ring-on-ring (RoR) test. Recently, systematic work was conducted by Stewart [10] also using RoR on a variety of PZT materials including PZT-5A and -5H from Morgan Matroc, and the strength values of two materials were found to be 88 and 82 MPa, respectively.

The Weibull moduli of 5A4E and 5H4E in this study were in the range of 16 to18 and 8 to 20, respectively, while those of Morgan PZT-5A and -5H were 9.3 and 19.3, respectively. It appears that the strength distribution of PZT-5A has more spread than that of 5A4E, while the data of both PZT-5H and 5H4E has similar scattering.

The differences in the reported values of strength may be attributed to the strength-size scaling as well as potential differences in compositions and microstructures (due to the different sources). If the failure of PZT disk is dominated by surface-type flaws, then failure stress can be scaled according to the following

$$S_B = \left(\frac{k_{sA} A_A}{k_{sB} A_B} \right)^{1/m} S_A \qquad (1)$$

where S_A and S_B are the failure stresses, k_{sA} and k_{sB} are the loading factors, and A_A and A_B are the total surfaces under tension for the two specimens (or components) of interest, respectively. With this respect, both the loading surface and the loading factor appear to be higher in the RoR configuration used in the above mentioned work [9, 10] than those of the current BoR configuration, and they have combined and contributed to the reduction of strength.

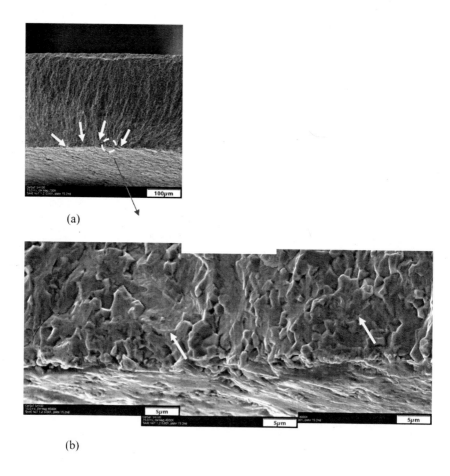

(a)

(b)

Fig. 4 Fracture surface of 5A4E for E = -1.2 kV/mm and σ_f = 114.7 MPa (the images were taken with the fracture surface tilted). (a) cross section of fracture surface whose failure origin is delineated by the arrows, (b) the enlarged area into the failure origin where transgranular fracture prevailed.

Sheet-to-sheet differences in strengths were also observed here, particularly for 5H4E (Table 2). The characteristic strengths of Sheet 3 ranged between 99 to 106 MPa under the OC, which are significantly lower than those of Sheet 1 under same condition (135 to 141 MPa). The Weibull modulus also varied correspondingly with 8 to 11 for Sheet 3, and 17 to 20 for Sheet 1. Because the two sheets were purchased in different periods of this project, it was suspected that processing parameters could have changed, which influenced the measured strength. That can be partly seen from the fractography results, especially in the case of 5H4E Sheet 3 where the intergranular fracture was revealed to prevail on the fracture surface. Although the fractographic study is still ongoing now for 5H4E Sheet 1, those on the specimens from 5H4E Sheet 1 exhibited the substantial transgranular fracture. It remains unknown at this point whether the intergranular fracture mode is a cause or an effect of the strength reduction.

Nevertheless, the strengths of both 5A4E and 5H4E consistently decreased under the negative E and increased under the positive E (Table 2). Strength reduction of PZT-5H was also observed by Stewart [10] with an applied electric field, but no strength enhancement was reported. In the tests of Zhoga and Shipeizman [9], an approximately symmetrical response of E effect on strength was observed. The biaxial strength of PZT-19 exhibited a slight increase under weak field and then significant decreased under high field disregarding the field direction. The symmetrical response of E effect was also observed by Fu and Zhang [11] on PZT-841 (American Piezo Ceramics) using three-point bending configuration; the bending strength continuously decreased when the electric field was applied both along and against the poling direction. The E_c of PZT-841 was 3.0 kV/mm and the applied field was less than the coercive field in the both directions and, therefore, the reduction in strength was attributed to the internal stress induced by the domain wall motion [11]. However, in the view of Zhoga and Shipeizman [9], the effect of domain wall would be related to the stress level; at lower stresses, the motion of domain wall may result in relaxation of structural overstress, while it could result in enhancement and formation of microcracks at high stresses. Zhoga and Shipeizman's comments are useful in understanding the related electro-mechanical effect, but it is clear that the different PZT systems and different E levels make our results hard to compare directly.

The vertical shift of the confidence ring of 5A4E (Fig. 3) suggests that the same flaw was responsible for the failure under both mechanical and electro-mechanical loading (i.e. no significant change in Weibull modulus). That is similar to observations by Yamashita et al. [12], and Zhoga and Shipeizman [9]. On the other hand, the diagonal shift (down and left in −E and up and right in +E) of confidence ring of 5H4E may signify different strength-limited flaw populations in the respective condition. Generally, a negative electrical field made the strength more scattered and a positive electrical field narrowed it. It is worthwhile to note that the identification of the strength-limited flaw based on the magnitude of Weibull modulus may be misleading somehow because its analysis assumes the material behaves linearly elastically up to fracture, and that may not be the case for some PZT materials where the localized inelastic phenomena dominates.

Given that the response of a surface processing flaw eventually dominates the fracture, one can see that the material would be strengthened under a positive coercive field due to a larger number of switchable domains and weakened under a negative field because of a fewer number of switchable domains. The details as regarding the role of crack tip domain switching on the toughening can be found in the reference [13].

CONCLUSIONS

Based on the above experimental results under BoR loading:

1. A single-layer plate of poled PZT appears weaker under a negative electric field and stronger under a positive electric field.

2. Characteristic strength is influenced by the electric field for the tested PZT materials, decreasing with a negative electric field and increasing with a positive field.

3. Weibull modulus is influenced by the electric field in different ways. For 5H4E, the Weibull modulus becomes lower under the negative field and higher under the positive. For 5A4E, it is barely affected.

4. Surface-type processing flaws appear to have limited the strength of these PZT materials.

ACKNOWLEDGEMENTS

This research was sponsored by the U.S. DOE, Office of FreedomCAR and Vehicle Technologies, as a part of the Heavy Vehicle Propulsion System Materials Program, under contract DE-AC05-00OR22725 with UT-Battelle, LLC. The work was supported in part by an appointment to the Oak Ridge National Laboratory Postdoctoral Research Associates Program, sponsored by the U.S. DOE and administered by the Oak Ridge Institute for Science and Education. Authors would like to thank K. E. Johanns, M. K. Ferber, H.-T., Lin and S. B. Waters. Lastly, the authors thank B. B. Hickey, P. F. Becher and C.-H. Hsueh for reviewing the manuscript.

REFERENCES

[1] Winzer, S. R., Shankar, N., and Ritter, A. P. Designing cofired multilayer electrostrictive actuators for reliability, J. Am. Ceram. Soc., 72 (12), 2246-57 (1989)

[2] Uchino, K., Furtuta, A., Destruction mechanism of multilayer ceramic actuators, Proc. ISAF 1992, Greenville, South Carolina, 195-198 (1992)

[3] Joshi, U., Kalish, Y., Savonen, C., Venugopal, V., and Henein, N., 2. Materials for fuel systems, Heavy Vehicle Propulsion Materials FY 2003 Progress Report, 7-14, (2003)

[4] Supancic, P., Wang, Z., Harrer, W., Reichmann, Danzer, R., Strength and fractography of piezoceramic multilayer stacks, Key Eng. Mater., 290 , 46-53 (2005)

[5] Shetty, D. K., Rosenfield, A. R., McGuire, P., Bansal, G. K., and Duckworth, W. H., Biaxial flexure tests for ceramics, Ceramic Bulletin, 59(12), 1193-1197 (1980)

[6] ASTM Standard C1239-00, Standard Practice for Reporting Uniaxial Strength Data and Estimating Weibull Distribution Parameters for Advanced Ceramics, Volume 15.01, March 2002

[7] ASTM Standard C1322-96a Standard Practice for Fractography and Characterization of Fracture Origins in Advanced Ceramics, Volume 15.01, March 2002

[8] Jaffe, H., Berlincourt, D. A., Piezoelectric transducer materials, Proc. IEEE, 53(10), 1372-1386 (1965)

[9] Zhoga, L.V., Shpeizman, V.V., Failure of ferroelectric ceramics in electric and mechanical fields, Sov. Phys. Solid State, 34, 2578–2583 (1992)

[10] Cain, M. G., Stewart, M. and Gee, M. G., Mechanical and electric strength measurements for piezoelectric ceramics: Technical measurement notes. NPL REPORT CMMT (A), 99, March, 1998.

[11] Fu, R., Zhang, T.Y., Influence of temperature and electric field on the bending strength of lead zirconate titanate ceramics, Acta Mater., 48, 1729–1740 (2000)

[12] Yamashita, K., Koumoto, K., Yanagida, H., Analogy between mechanical and dielectric strength distributions for BaTiO3 ceramics. Commun. of Am. Cerm. Soc., C-31-C-33, Feb. 1984

[13] Mehta, K., Virkar, A. V., Fracture mechanisms in ferroelectric- ferroelastic lead zirconate titanate (Zr:Ti = 0.54:0.46) ceramics, J. Am. Ceram. Soc., 73(3), 567-574 (1990)

PIEZOELECTRIC CERAMIC FIBER COMPOSITES FOR ENERGY HARVESTING AND ACTIVE STRUCTURAL CONTROL

R. B. Cass, Farhad Mohammadi
Advanced Cerametrics, Inc.
Lambertville, NJ, USA

ABSTRACT:
Advanced Cerametrics, Inc. (ACI) continues to optimize its Viscous Suspension Spinning Process to form continuous lengths of piezoelectric ceramic fibers to design and build waste energy scavenging systems. These ceramic fibers are made into polymer composites that can harvest waste energy to create functional amounts of power or to actively drive structures to reduce vibration or use ambient mechanical energy (or an external power supply) to morph the structures. Recent results include, for example, the production of 125 mW of continuous power at 22 Hz of vibration (the vibration source displacement and force were 1.4 mm and 1.2 lb of force, respectively). This is sufficient power from the ceramic fiber composites to run many conventional electronic systems, eliminating the need for batteries, fuel cells or power connections. Demonstrated results show that this ceramic fiber technology to be a viable, green, inexpensive future source of power.

BACKGROUND/INTRODUCTION: ENERGY HARVESTING

Energy harvesting (EH), sometimes referred to as energy scavenging, has gained tremendous attention as a means to lessen or eliminate the need for battery power. Novel approaches have been available to generate power by utilizing energy from human and environmental sources. Given battery limitations, wider adoption of EH is coming but requires a clever combination of design skills. A multi-discipline approach that leverages knowledge in several fields of engineering is required, including electrical, mechanical, material, and process. EH itself is not new. Consider hand-cranked radios, flashlights powered by shaking, windmill farms, and solar energy. What is new is the application of EH for ultra low power embedded electronics. The convergence of high charge piezoelectric ceramics and Advanced Cerametrics' ceramic fiber composite process technology has enabled the application of piezoelectric-powered systems for a wide array of electronic systems.

Piezoelectricity is the ability of certain crystalline materials to develop an electric charge proportional to a mechanical stress (direct effect), and a geometrical strain (deformation) proportional to an applied voltage (converse effect)[1]. The converse effect is used in piezoelectric actuators and makes possible quartz watches, micropositioners, ultrasonic cleaners, and an enormous number of other important products. A mechanical stress applied on a piezoelectric material creates an electric charge. Piezoceramics will give off an electric pulse even when the applied pressure is as small as sound pressure. This is called the direct piezoelectric effect and is used in sensor applications such as microphones, undersea sound detecting devices, pressure transducers, etc. It is the direct effect, which is used in piezoelectric energy harvesting applications. Schematics of the direct and converse effects are shown in Figure 1. The generated charge is proportional to the external force pressure and disappears when the pressure is withdrawn. Some examples of the piezoelectric effects in a sampling of product developments are shown in Table I.

ULTRA LOW POWER ELECTRONICS

The science of piezoelectric devices is fairly well understood in the engineering world, but their application remains a nascent field rich with possibilities. The emergence of piezoelectric ceramic fiber composite transducers that offer increasing deliverable power, combined with electronic components that measure performance in nanowatts, is opening a wide range of new product and services. The need for Extreme Life Span Power Supplies (ELSPS) for numerous electronic systems and devices is fueling extensive research, development, and growth. Piezoelectric ceramic fibers, given their unique properties of flexibility, light weight and higher output per pound of material, offer the greatest potential for enabling the wide-scale deployment of self-powered systems. ACI's current state of the art in this technology is the generation and storage of functional amounts of power from forces as low as 100 milli-G's. Improved electronics and fiber composite design have raised the piezo conversion efficiency significantly over the last year.

PIEZOELECTRIC CERAMIC FIBERS

Conventional piezoelectric ceramic materials are rigid, heavy, and produced in block form. ACI's low-cost piezo fiber forming technology, termed the Viscose Suspension Spinning Process (VSSP), can produce fibers that range in diameter from 10-250 μm (Figure 2).[2, 3] When formed into user defined (shaped) composites, the ceramic fibers possess all the desirable properties of ceramics (electrical, thermal, chemical) but eliminate the detrimental characteristics (brittleness, weight). The VSSP generates fibers with more efficient energy conversion than traditional bulk ceramics due to their, essentially, large length to area ratio ($V = g_{33}F L/A$, where V is piezoelectric generated output voltage, g_{33} is piezoelectric voltage coefficient, L and A are fiber length and cross-section area, respectively[4]). To put this into perspective, mechanical to electrical transduction efficiency can reach 70-75%, where vibrations can be harvested 24 hours per day. The VSSP can produce fibers from almost any ceramic material.

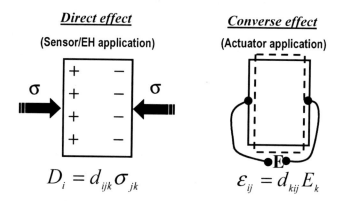

$$D_i = d_{ijk}\sigma_{jk}$$

$$\varepsilon_{ij} = d_{kij}E_k$$

Figure 1. Schematics representing piezoelectric direct and converse effects.

Table I. Piezoelectric effects in various applications.

Application	Direct Piezo Effect	Inverse Piezo Effect
Energy Harvesting		
Battery Replacement	√	
Vibration Reduction		√
Reusable Energy Supply	√	
Condition Monitoring Sensors	√	√
Sporting Goods		
Smart Skis and Tennis	√	√
Racquets	√	√
Golf Equipment	√	√
Baseball bats, hockey sticks	√	√
Medical Devices		
Ultrasound Imaging	√	√
Self-Powered Pacemaker	√	
Defense		
Vibration Suppression		√
Active Structure Control	√	√
Sonar, Hydrophones	√	√
Ice Thickness Sensing	√	
Acoustical Devices		
Audio Reproduction	√	
Electronic Equipment	√	
Acoustic Suppression		√
Sensing		
Level & Weight Sensors	√	√
Non Destructive Testing	√	√
Smart Bearings	√	√

PIEZO POWER GENERATION

Piezoelectric Fiber Composites (PFC) open the door for an array of energy harvesting applications. The fiber can recover (harvest) waste energy from mechanical forces such as motion, vibration, and compression (strain). With simple, low-cost analog circuits, the piezo power can be converted, stored, and regulated as a direct replacement for batteries. A typical single, PFC can easily generate voltages in the range of 80 V_{p-p} from vibration. A typical single, PFCB (bimorph) can easily generate voltages in the range of 500 V_{p-p} with some forms reaching outputs of 4000 V_{p-p}. Figure 3 shows the power generation capabilities of a PFCB transducer. The vibration frequency was 22 Hz and the source displacement and force were 1.4 mm and 1.2 lb of force, respectively. In this experiment, it only took 3 seconds to fill a 400 μF capacitor to 50 V. One minute of vibration at 125 mW can store 7.5 J of energy in a capacitor bank that can run some devices for hours. The generated energy was stored in a capacitor bank and the time, t, to charge the capacitor was measured. The voltage, V, in the capacitors was 50 V and the power, p, was calculated via $P=(1/2\ CV^2)/t$. The variation of capacitance, C, had minimal effect on the power. Energy sufficient to power wireless systems for sensing and control of equipment, toll transponders, appliances, medical devices, buildings, and other infrastructure elements has been demonstrated.

Power output is scalable by combining two or more piezo elements in series or parallel, depending on the application. The composite fibers can be molded into unlimited user defined shapes and are both flexible and motion sensitive. The fibers are typically placed where there is a rich source of mechanical movement or waste energy. Another piezoelectric transducer type for energy harvesting is called piezoelectric multilayer composites (PMC). The properties of a typical PMC are shown in Table II. In PMC's, the generated output voltage and power increases with increasing transducer thickness and decreasing fiber diameter, making small diameter piezoelectric fibers very attractive for energy harvesting.

Table II. Typical properties of ACI's PMC (1-3 type) composites

Fiber type	PZT-EC65 (EDO)
Curie temperature (°C)	350
Fiber volume fraction (%)	30
Fiber diameter (μm)	13
Piezoelectric charge coefficient, d_{33} (pC/N)	375
Dielectric constant	700
Dielectric loss (%)	2
Coupling coefficient, k_t	0.65
Compliance, $S_{33}(D)$	2.93×10^{-12}
Compliance, $S_{33}(E)$	5.07×10^{-12}
Elastic constant $Y_{33}(D)$	3.42×10^{11}
Elastic constant $Y_{33}(E)$	1.97×10^{11}

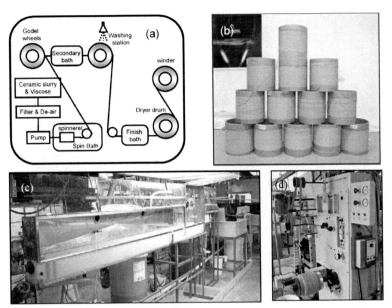

Figure 2. Viscose Suspension Spinning Process (VSSP): (a) Schematic of VSSP, (b) spools of green PZT fiber, (c) VSSP production line, and (d) automatic fiber winding machine.

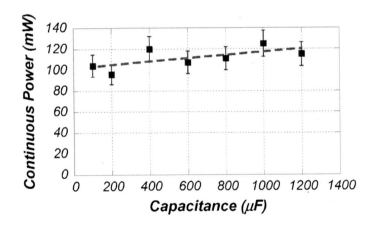

Figure 3. Power generation capability of a PFCB vibrating at its resonance frequency of 22 Hz. The vibration amplitude and force was 1.4 mm and 1.2 lbf, respectively.

EXAMPLE APPLICATION: WIRELESS SENSOR NETWORKS

Sensors that measure everything from process temperatures, to system pressures, to machine vibrations have been historically expensive to deploy in manufacturing and industrial environments. The sensors require expensive wiring and are expensive to service. With the emergence of the new Zigbee standard, based on IEEE 802.15.4, the availability of large, low-cost, low-power wireless sensor networks (WSNs) that are self-managed has become a reality. Sensors, signal conditioners, controllers, and RF transceivers continue to become smaller, lower power, and highly integrated. The combination of wireless networking, intelligent sensors, and distributed computing has created a new paradigm for monitoring the health of machines, people, buildings, and the environment.

A low-cost, renewable energy source is critical to ubiquitous deployment of WSNs. After all, who wants to have to change thousands of batteries and then find a way to dispose of them? New piezoelectric fiber-based energy harvesters, will in some cases, obviate the need for batteries in the WSNs. In other cases the harvesting technology can be used to recharge batteries to enhance service life. The power comes from the vibration or other ambient mechanical energy sources of the system being monitored. Piezo fiber-based products will require no maintenance, significantly reduce the life cycle costs, and improve the overall quality of industrial and machine control systems. This is a truly 'green' technology.

Figure 4 shows an example of the PZT ceramic fiber acting as an energy harvester to convert waste mechanical energy into a self-sustaining power source for a Zigbee wireless sensor node. The piezoelectric fiber captures the energy generated by the structure's vibration, compression, or flexure. The resulting energy (current) is used to charge up a storage circuit that then provides the necessary power level for the sensor node electronics. In this example, energy is harvested by the vibration of PZT fiber composites. The energy is converted and stored in a low-leakage charge circuit until a certain threshold voltage is reached. Once the threshold is reached, the regulated power is allowed to flow for a sufficient period to power the Zigbee controller and RF transceiver.

ADDITIONAL APPLICATIONS
Wireless Transponders in Automobiles

There are numerous low-power automotive applications where batteries and wires can be replaced. For example, wireless transponders are used in some automobiles for payment of tolls. These are, problematically, powered by batteries, but recent experiments at ACI have shown that these can be easily powered via a piezoelectric fiber composite taking advantage of car vibrations. There are several sources of vibration associated with a moving vehicle. These include: (1) engine vibration, (2) vehicle acceleration/deceleration, (3) air turbulence, (4) vehicle turning, and (5) road condition. In this application, an energy of 1.5 mJ was required for one wireless transmission. Three different cars were selected that include a truck, a SUV, and a compact sedan in a road test. In all these cases, the required energy for one transmission was produced in less than 70 seconds (Figure 5). Even though, this amount of energy can replace the currently used batteries, work is underway to reduce charging time via transducer design modifications.

Lighting

PFC's can convert mechanical energy directly into light energy with no intervening electronics. By harvesting energy from ambient vibrations, PFC's can provide electroluminescent lighting on bridge decks, digital signage, buoys, and other low-power lighting loads.

Smart Structures

PFC's also offer solutions for vibration damping and structural morphing. To enable self-adjusting systems, a smart structure containing a PFC senses a change in motion. The motion produces an electrical signal that can be sent to a control processor that measures the magnitude of the change in motion and returns an amplified signal that either stiffens or relaxes the PFC actuators/sensors. Figure 6 demonstrates the application of PFC transducers in powering various wireless transmitters.

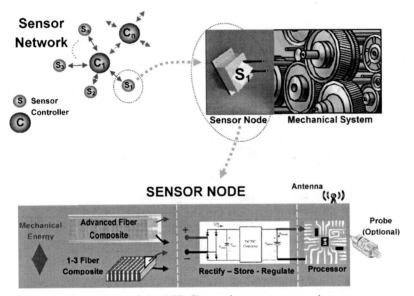

Figure 4. An example of the PZT fiber acting as an energy harvester to convert waste mechanical energy into a self-sustaining power source for a Zigbee wireless sensor node.

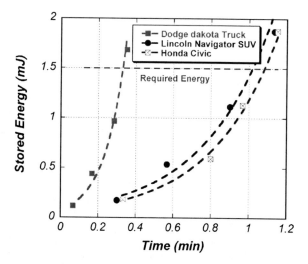

Figure 5. Time required to produce and store sufficient energy for one wireless transmission, when driving various vehicles in a road test.

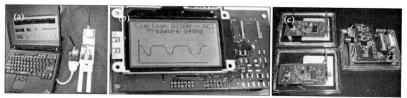

Figure 6. Examples of piezoelectric fiber powered wireless sensing systems: (a) transmitter/receiver, (b) pressure sensor receiver, (c) two encoded transmitters and a receiver.

SUMMARY

There are an emerging number of new and unique products coming to market that are limiting or obviating the need for batteries. Piezoelectric ceramic fiber technology provides a unique solution for EH, active structural control, and self-powered electronic systems. PFCB's made by ACI produced 125 mW of continuous power and have demonstrated durability exceeding 1 billion cycles. By combining fiber composites with low-cost electronics and packaging, a new era of ultra low power products and applications is dawning. New market solutions are emerging that offer an ELSPS Factor for low power applications that is battery free.

REFERENCES

[1] W.G. Cady, Piezoelectricity: An Introduction to the Theory and Applications of Electromechanical Phenomena in Crystals, New York: McGraw Hill, (1946).

[2] R.B. Cass, R.R. Loh, and T.C. Allen, "Method for Producing Refractory Filaments," *United States Patent*, No. 5,827,797, (1998).

[3] J.D. French and R.B. Cass, "Developing Innovative Ceramic Fibers," *American Ceramic Society Bulletin*, PP. 61-65, (1998).

[4] F. Mohammadi, A. Khan, R.B. Cass, "Power Generation from Piezoelectric Lead Zirconate Titanate Fiber Composites," *Materials Research Society Symposium*, V. 736, D5.5.1, (2003).

DIELECTRIC PROPERTY OF RESIN-BASED COMPOSITES DISPERSING CERAMIC FILLER PARTICLES

Jun Furuhashi, Ken-ichi Kakimoto
Depertment of Materials Science and Engineering, Graduate School of Engineering, Nagoya Institute of Technology
Gokiso-cho, showa-ku, Nagoya 466-8555, Japan

Toshiaki Yagi, Hidetoshi Ogawa, and Minoru Aki
Advanced Materials Laboratories, Otsuka Chemical Co., Ltd.
463 Kagasuno, Kawauchi-cho, Tokushima 771-0193, Japan

ABSTRACT

Resin-based composites dispersing ceramic filler particles have been prepared by using three different press-form methods of "tape-casting", "slurry-casting" or "press-casting", to compare the relationship between microstructure and dielectric property. Although the pore volume involved in the composites showed similar values despite casting methods, the median pore-size diameter demonstrated 0.5, 67.3 and 50.1 μm for the composites formed by "tape-casting", "slurry-casting" and "press-casting", respectively. As a result, they also showed different dielectric constant (at 100 kHz) of 50.4, 25.8 and 26.3, respectively.

INTRODUCTION

Electric devices such as antenna modulus working at high operating frequencies require dielectric materials with adjustable their dielectric constants.[1,2] Resin-matrix composites dispersing $BaTiO_3$-based ceramic filler particles have been attracted for these application, because their dielectric constants can be controlled by changing the filer concentration, shape and size.[3,4] However, the relation between dielectric property and microstructure is not well understood.

In this study, $(Ba_xCa_{0.1}Sr_{0.9-x})(Ti_{0.9}Zr_{0.1})O_3$ (x=0.4~0.8, abbreviated to 10xBT) was adopted as a ceramic filler material. After decision of the best composition for dielectric property in the form of the sintered ceramics, its ceramic powder for usage as the filler was synthesized by a molten salt method. Three kinds of resin-matrix composites were prepared by different press-form techniques. This paper compares their dielectric properties and microstructures to understand the effect of porosity on the polarization under electric fields.

Reagent-grade $BaCO_3$, $CaCO_3$, $SrCO_3$, TiO_2 and ZrO_2 were weighed and mixed in ethanol for 24h by ball-milling to prepare the nominal compositions of 10xBT. The dried mixture was calcinated in air at 1150°C for 2h, followed by pressing into a disk shape. The pressed sample was sintered in air at 1350°C for 2h, then subjected to dielectric measurement to select the best composition with the highest dielectric constant and the lowest dielectric loss at room temperature.

In this test, $(Ba_{0.4}Ca_{0.1}Sr_{0.5})(Ti_{0.9}Zr_{0.1})O_3$ (4BT) was selected as a ceramic filler composition suitable for resin-matrix composites. The filler was prepared by using a molten salt method. The slurry of 4BT was mixed with KCl in weight ratio of 1:1. The dried mixture was heated at 1250°C for 2h. After heat-treatment, KCl residue was removed from the product by repeated washing with hot distilled water. The product was dried and sieved to form the ceramic filler powder. The derived powder was then inserted to a container filled with a paraffin resin (melting point = 62°C, Wako pure chemical industries) that was kept at 90°C. The volume ratio of ceramic filler was 40 vol% against resin in the mixture. The ceramic/resin slurry was kept stirring for well dispersion within the mixture before composites were prepared at room temperature. The composites were formed by three kinds of different press-form techniques, as illustrated in Figure 1. "Tape-casting" method fabricated thick films of 700μm by using a doctor blade. "Slurry-casting" method produced composite films in a stainless mold by uni-axially pressing with hand power. On the other hand, 20 MPa was applied for the casting of the composites in "press-casting" method. The sample was formed into disks of 12 mm in diameter and 5 mm in thickness. The composites were cooled for 5 min to harden.

The crystalline phase and microstructure was characterized by X-ray diffraction (XRD) and scanning electron microscopy (SEM), respectively. The temperature dependence of dielectric property was evaluated in the temperature range from -40 to 30°C by using a LCR meter and an environmental chamber. The pore size distribution in the composite was measured by using a mercury porosity meter.

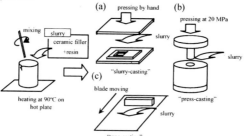

Fig. 1. Press-form methods.

RESULTS AND DISCUSSION

Figure 2 shows the XRD patterns of 10xBT ceramics. All samples showed a perovskite single phase and no secondly phase, indicating the formation of a solid solution. The every peak shifted toward higher angles with decreasing Ba content since Ba^{2+} is larger than Sr^{2+}. The temperature dependence of dielectric constant and loss tangent at 100 kHz for 10xBT ceramics is shown in Figure 3. The dielectric anomaly corresponding to the phase transformation between tetragonal and cubic was observed at the specific temperature. The temperature of maximum dielectric constant (T_{max}) shifted toward lower with decreasing Ba content, and was assigned to be 82 and 43 °C for 8BT and 7BT, respectively. In contrast, 6BT, 5BT and 4BT did not show obvious T_{max} peaks in the measured temperature range. This tendency was also observed in the measurement of dielectric loss tangent. Of all the candidates, 4BT was selected as a ceramic filler composition, because of its relatively large dielectric constant and low loss tangent (ε_r=1700, $\tan\delta$ =3.4×10⁻⁴ at 25°C). A single phase 4BT ceramic powder was also obtained by the molten salt method (Figure 2). SEM confirmed that the obtained 4BT powder is an isotropic granular shape and showed relatively uniform size distribution with an average particle size around 1 μm.

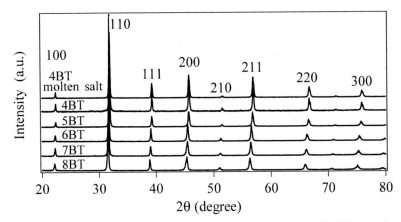

Figure 2. XRD patterns of $(Ba_xCa_{0.1}Sr_{1-x})(Ti_{0.9}Zr_{0.1})O_3$ (x=0.4~0.8, 10xBT) ceramics.

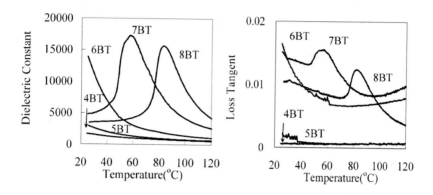

Figure 3. Temperature dependence of the dielectric property for $(Ba_xCa_{0.1}Sr_{1-x})(Ti_{0.9}Zr_{0.1})O_3$ (x=0.4~0.8, 10xBT) ceramics at 100kHz.

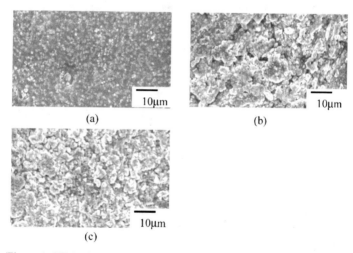

Figure 4. SEM micrographs of the composites prepared by (a) tape-casting, (b) press-casting, and (c) slurry-casting.

The SEM micrographs of the three kinds of ceramic/resin composites are shown in Figure 4. The composites prepared by "slurry-casting" and "press-casting" contained a lot of pores with their size is 10µm or more. It seems that air was easily trapped into the composite when the pressing was performed, because air could not be escaped outward from the mold in their forming step. In contrast, the composite synthesized by "tape-casting" showed almost no pore microscopically, because the sheer stress induced by moving of a doctor blade dominated and large pores were reduced to small pores or pulled out toward the side direction of the composite during forming process.

Figure 5 shows the pore size distribution of the composites. The median diameter of pore size distribution for "tape-casting", "slurry-casting" and "press-casting" was 0.5, 67.3 and 50.1 µm, respectively. The pore diameter of "tape-casting" was much smaller than the others. This result is in good agreement with the SEM observation. On the other hand, the total pore volumes of the composites were almost no dependence on the casting methods and were calculated to 15.2, 16.2 and 18.5% for "tape-casting", "slurry-casting" and "press-casting", respectively.

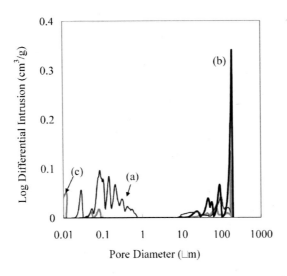

Figure 5. Pore size distribution of the composites prepared by (a) tape-casting, (b) press-casting, and (c) slurry-casting.

Figure 6 and Table 1 present the temperature property of dielectric constant measured for the specimens prepared by three kinds of the press-form methods. At room temperature, the dielectric constant for the composites prepared by "tape-casting", "slurry-casting" and "press-casting" was 50.4, 25.8 and 26.3, respectively. The dielectric constant of the composite prepared by "tape-casting" showed nearly twice value than those of the composites prepared by "slurry-casting" and "press-casting", although loss tangent was in the order of 10^{-3} for all the composites in the temperature range measured. It is generally accepted that the dielectric constant of ceramics is changeable by inner stress during processing.[5] In the present study, however, pressing stress was induced to the resin matrix with low dielectric constant in most case, because resin is mechanically soft compared with ceramic powder with high dielectric constant. Therefore, the observed difference in the dielectric constant among specimens seems to have been closely correlated with porosity rather than stress. It is considered that local large pore prevented from applying electric field to ceramic filler directly, then limited the polarization of ceramic fillers under electric fields.

The temperature coefficient of dielectric constant, τ_ε, in the range from -40 to 25°C was also determined from the curves shown in Figure 6, and listed in Table 1. τ_ε of the composite prepared by "tape-casting" demonstrated the highest of all the composites. This result may reflect that the inherent dielectric property of 4BT could appear in the composite prepared by "tape-casting" rather than the other pressing methods.

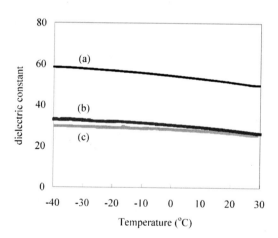

Figure 6. Temperature dependence of the dielectric constant for the composites prepared by (a) tape-casting, (b) press-casting, and (c) slurry-casting at 100kHz.

Table 1 dielectric property and   median pore diameter

	dielectric constant at 25°C			τ_c	median pore
	1kHz	10kHz	100kHz	(ppm/°C)	diamiter (μm)
cf. 4BT ceramics	1741	1744	1744	-183	
cf. resin (tape-casting)	3.1	3.4	3.6	0	
composite (slurry-casting)	26.4	26.0	25.0	-0.67	67.3
composite (press-casting)	26.8	27.0	27.1	-0.81	50.1
composite (tape-casting)	51.2	50.8	50.4	-1.16	0.5

CONCLUSION

$(Ba_xCa_{0.1}Sr_{1-x})(Ti_{0.9}Zr_{0.1})O_3$ (x=0.4~0.8) ceramics was synthesized and their dielectric property was evaluated. The x=0.4 sample (4BT) was selected as a ceramic filler suitable for resin-based composite, because it has relatively large dielectric constant of 1700 and low loss tangent of 3.4×10^{-4} at 25°C. The 4BT particles synthesized by molten salt method showed uniform size distribution and isotropic granular shape.

Next, the resin-based composites including the 4BT filler were synthesized by three different press-form methods. Of all materials, the composite prepared by "tape-casting" showed the highest dielectric constant of 50.4 as well as low loss tangent of the order of 10^{-3} at 100 kHz. It was found that smaller pore-size distribution worked effectively to enhance the dielectric constant of the resin-based composites dispersing ceramic filler particles.

REFERENCES

[1] D-H. Kuo, C-C. Lai, and T-Y. Su, "Dielectric Behavior of Nb_2O_5-doped TiO_2/epoxy Thick Films," *Ceram. Int.*, **30**, 2177-81(2004).

[2] F. Baba, "Koushuuhayou Koubunshizairyou no Kaihatsu to Ouyou (in Japanese)," CMC Publishing Co.,Ltd. (Japan), pp. 3-127 (1999).

[3] S. Che, I. Kanada, and N. Sakamoto, "Dielectric Property of Spherical Oxide Powder and Its Ceramic-Polymer Composite," *Jpn. J. Appl. phys.*, **44**, 7107-10 (2005).

[4] D-H Kuo, C-C Lai, T-Y Su, W-K Wang, and B-Y Lin, "Dielectric Properties of Three Ceramic/Epoxy Composites," *Mater. Chem. and Phys.*, **85**, 201-06 (2004).

[5] L. Szymczak, Z. Ujma, J. Handerek, and J. Kapusta, "Sintering Effects on Dielectric Properties of $(Ba,Sr)TiO_3$ Ceramics," *Ceram. Int.*, **30**, 1003-08 (2004).

THE EFFECT OF SINTERING CONDITIONS AND DOPANTS ON THE DIELECTRIC
LOSS OF THE GIANT DIELECTRIC CONSTANT PEROVSKITE $CaCu_3Ti_4O_{12}$

Barry A. Bender and Ming-Jen Pan
Naval Research Lab
Code 6351
Washington, DC 20375

ABSTRACT
    The effect of dopants and oxygen in sintering on the dielectric loss of $CaCu_3Ti_4O_{12}$
(CCTO) has been explored. Sintering in oxygen affected the kinetics of abnormal grain growth
which influenced the resultant microstructures and dielectric properties. Lower sensitivity of loss
to temperature was measured for the oxygen-sintered materials. Doping with 0.2 w/o hafnia and
0.25 w/o titania also led to CCTO ceramics with lower loss temperature sensitivity. Doping with
0.3 w/o CaO led to larger grain growth and substantial increases in dielectric constant and loss.

INTRODUCTION
    The Navy is engaged in research to develop the all-electric ship. Through its power
electronic programs the Navy has determined that in their state-of-the-art power converters that
the size of the filter capacitors is a limiting factor in their size reduction. Miniaturization of the
capacitors demands greater volumetric efficiency of capacitance through materials which have
higher permittivity. These capacitors also have to be stable over a wide range of temperatures
and operating voltages. Commercial dielectric ceramics such as ferroelectric $BaTiO_3$'s typically
fulfill two of the above conditions but not all three. However, recent electrical property
measurements of the perovskite $CaCu_3Ti_4O_{12}$ show that it has the potential to satisfy all of the
above requirements. Subramanian et al.[1] were the first to measure CCTO's dielectric properties
and they found that polycrystalline $CaCu_3Ti_4O_{12}$ possessed a dielectric constant of 12,000 (room
temperature- 1 kHz) that exhibited little temperature dependence from zero to 200°C.
Permittivity measurements on single crystal CCTO showed a giant dielectric constant as high as
80,000.[2] As a result, $CaCu_3Ti_4O_{12}$ ceramics have generated great interest because of their
potential use in miniaturization of commercial electronic devices.
    For this material to be used commercially its dielectric loss properties have to be
improved. Room temperature values as low as 0.05 (1 kHz) have been reported.[3, 4] However, the
dielectric loss of these CCTO ceramics is sensitive to temperature as dielectric loss-temperature
curves start to warp up at temperatures as low as 40°C leading to losses that quickly exceed 0.10
before 60°C is reached.[3,5-7] Before the dielectric loss properties can be improved the nature of
the giant permittivity of $CaCu_3Ti_4O_{12}$ has to be better understood. Researchers believe that
CCTO's giant permittivity is extrinsic in nature and is the result of the formation of internal
capacitive barrier layers.[1,8,9] Researchers believe that insulating surfaces form during processing
at the grain boundaries of semiconducting grains. This creates an electronically heterogeneous
material comparable to internal barrier layer capacitors (IBLCs). Chung et al.[10] using
microelectrodes showed conclusively that a large electrostatic potential barrier exists at the grain
boundaries in CCTO with n-type semiconducting grains separated by an insulative grain
boundary region. However, the exact mechanisms responsible for the semiconducting nature of
the grains and the nature of the grain boundary region are still up for scientific debate. The
electrical measurements of CCTO by Adams et al.[11] and Zang et al.[12] show that the electrostatic

potential barrier can be best described using a double Schottky barrier (DSB) model. DSB's play an important role in the properties of varistors and PTCR's (positive temperature coefficient resistors). A myriad of research has been done showing the importance of the role of oxygen and dopants on the properties of varistors and PTCRs.[13-15] Research on lead iron niobate (PFN), which exhibits PTCR behavior, has shown that sintering in oxygen and the addition of dopants can lower its loss.[16, 17] This paper explores the effect of oxygen in sintering and the dopants of hafnia, titania, and calcia on the dielectric loss and permittivity of various $CaCu_3Ti_4O_{12}$ ceramics.

EXPERIMENTAL PROCEDURE

$CaCu_3Ti_4O_{12}$ was prepared using ceramic solid state reaction processing techniques. Stoichiometric amounts of $CaCO_3$ (99.98%), CuO (99.5%) and TiO_2 (99.5%) were mixed by blending the precursor powders into a purified water solution containing a dispersant (Tamol 901) and a surfactant (Triton CF-10). The resultant slurries were then attrition-milled for 1 h and dried at 90°C. The standard processed powder, STD, was calcined at 900°C for 4 h and then 945°C for 4 h. After the final calcination the STD powders were attrition-milled for 1 h to produce finer powders. The hafnia-doped powder, HSTD, was fabricated by mixing 0.2 w/o HfO_2 (99.9%) with the calcined STD powder. The calcia-doped powder, CSTD, and titania-doped powder (TSTD), were made by mixing 0.3 w/o $CaCO_3$ (99.9%) or 0.25 w/o TiO_2 (99.5%) with the calcined STD powder. A 2% PVA binder solution was mixed with the powders and they were sieved to eliminate large agglomerates. The dried powder was uniaxially pressed into discs typically 13 mm in diameter and 1 mm in thickness. The discs were then placed on platinum foil and sintered in air or flowing oxygen at 1100°C for various dwell times (2, 8, and 16 h).

Material characterization was done on the discs and powders after each processing step. XRD was used to monitor phase evolution for the various mixed powders and resultant discs. Microstructural characterization was done on the fracture surfaces using scanning electron microscopy (SEM). To measure the dielectric properties, sintered pellets were ground and polished to achieve flat and parallel surfaces onto which palladium-gold electrodes were sputtered. The capacitance and dielectric loss of each sample were measured as a function of temperature (-50 to 100 °C) and frequency (100 Hz to 100 KHz) using an integrated, computer-controlled system in combination with a Hewlett-Packard 4284A LCR meter.

RESULTS

The Effect of Oxygen in Sintering on Microstructure, Dielectric Loss and Permittivity

Sintering in oxygen of the STD powders at 1100°C for 2 h resulted in major differences in both microstructure and dielectric properties. Figure 1(a) shows that a bimodal distribution of grain sizes developed when the STD powder was sintered in air. The average grain size of the large grains was 22 microns. The volume fraction of large grains was 26%. A large percentage of the grains were isolated by themselves and were pore-free. The average grain size of the small grains was 1.3 microns. This matched well with the average grain size of the oxygen-sintered STD powder of 1.4 microns but no large grains formed in this CCTO (see Fig. 1(b)).

Substantial differences were also observed between the 2 samples in regards to dielectric properties. As shown in Fig. 2 substantial reductions in dielectric loss and permittivity resulted when the STD powder was sintered in oxygen. The dielectric loss dropped (1 kHz- 20°C) from 0.058 to 0.012. Its sensitivity to temperature was reduced too as the dielectric loss rose to only 0.02 at 75°C as compared to the dramatic increase to 0.38 for the air-sintered material. The reduction in loss was also accompanied by a reduction in permittivity from 17,400 to 2100.

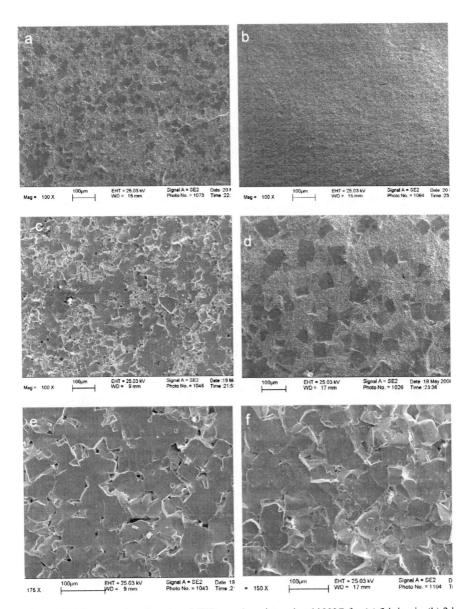

Fig. 1 SEM fractographs of undoped STD powders sintered at 1100°C for (a) 2 h in air, (b) 2 h in oxygen, (c) 8 h in air, (d) 8 h in oxygen, (e) 16 h in air, and (f) 16 h in oxygen.

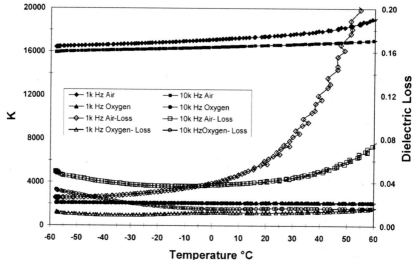

Fig. 2. Temperature dependence of dielectric constant and loss for STD powder sintered for 2 hours in either air or flowing oxygen at 1100°C.

Substantial differences in microstructure were again observed for the STD powder sintered in oxygen at 1100°C for 8 h. Figure 1(c) shows that the air-sintered CCTO consisted primarily of large grains (97% volume fraction) with an average grain size of 40 microns. However, the oxygen-sintered material developed a bimodal microstructure (Fig. 1(d)) where the volume fraction of large grains was only 35%. Their average grain size was 50 microns with the larger grains tending to be isolated from each other and more porous.

However, the dielectric properties of air and oxygen-sintered materials for 8 h were more comparable than the CCTO ceramics sintered for 2 h. The dielectric loss for both materials was 0.022 at 20°C. However, again the air-sintered material dielectric loss was more sensitive to temperature as its loss increased at 75°C to 0.08 (see Fig. 3) as compared to only 0.03 for its counterpart. Sintering in oxygen lead to a reduction in dielectric constant from 35,700 to 27,500.

After sintering the STD powder for 16 h in oxygen the microstructure and dielectric properties are comparable to the material sintered in air. Both microstructures consist of only large grains. The grains of the oxygen-sintered CCTO are larger (51 vs 38 microns) and more porous (see Fig. 1 (e and f)). The dielectric constant and loss of the oxygen-sintered CCTO are slightly larger (40,600 and 0.028 vs 40,000 and .018). The dielectric loss-temperature curve for the air-sintered material (see Fig. 4) shows a much flatter dielectric loss-temperature curve as compared to its counterpart sintered for 8 h but at 100°C the dielectric loss is still lower for the oxygen-sintered material (0.084 vs 0.100).

The Effect of Dopants on Microstructure, Dielectric Loss and Permittivity

Figures 5 and 6 show the microstructural and dielectric property differences between the HfO$_2$-doped and undoped STD sintered in air for 16 h. Doping with 0.2 w/o HfO$_2$ reduced the average grain size from 38 to 30 microns (Fig. 5(a & b)). The micrographs indicate that there is less porosity in the grains of the HSTD CCTO and this is reflected by the sample's higher relative density (94.6% vs 93.1%). Figure 6 shows that doping with hafnia slightly lowered the dielectric loss at room temperature from 0.018 to 0.016. It also shows that it substantially affected the temperature response of the dielectric loss as HSTD had a dielectric loss at 100 °C of only 0.053 as compared to the STD material whose loss practically doubled (0.100). Hafnia-doping also lowered the room temperature permittivity (1 kHz) from 40,000 to 30,800.

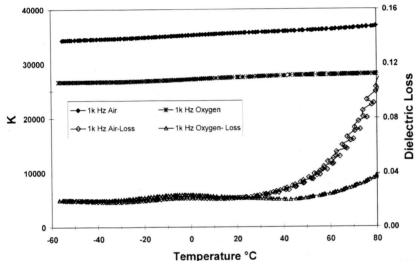

Fig. 3. Temperature dependence of dielectric constant and loss for STD powder sintered for 8 hours in either air or flowing oxygen at 1100°C.

Doping with 0.3 w/o CaO led to major differences in microstructure and dielectric properties. Figure 5 (c & d) shows that doping with calcia led to a 16% increase in grain size (44 vs 38 microns). Similar to the hafnia-doped CCTO there appears to be less porosity in the grains of the CSTD ceramic. Figure 7 shows that doping with CaO had a dramatic effect on dielectric loss as the dielectric loss at room temperature climbed from 0.018 for the STD CCTO to 0.087 for the CSTD ceramic. Calcia-doping also led to a striking increase in temperature sensitivity to loss as the dielectric loss rose to 0.60 at a temperature of only 75°C.

Figures 5 and 7 show some of the microstructural and dielectric property differences between the titania-doped and undoped CCTO. Doping with 0.25 w/o titania led to a small reduction in grain size (Fig. 5 e & f) from 38 to 35 microns (90% confidence level- 38 ±1.6 and 35 ±1.3). Also like the other studied doped-CCTO's it appears that doping lead to a decrease in the internal porosity of the grains. Figure 7 shows that doping with titania improved the dielectric loss of $CaCu_3Ti_4O_{12}$. Though there is a small increase in room temperature dielectric loss from 0.018 to 0.020 the TANT's sensitivity in regards of dielectric loss to temperature is substantially lowered as its loss value is only 0.048 at 100°C and it doesn't reach 0.10 until a temperature of 132°C.

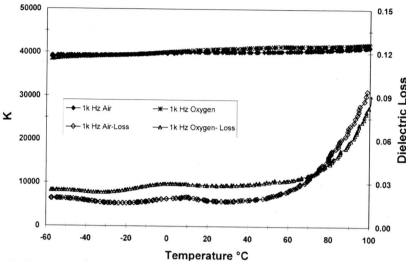

Fig. 4. Temperature dependence of dielectric constant and loss for STD powder sintered for 16 hours in either air or flowing oxygen at 1100°C.

DISCUSSION

The Effect of Oxygen in Sintering on Microstructure and Dielectric Loss

Various facets of the mechanisms that lead to the formation of Schottky barriers in CaCu$_3$Ti$_4$O$_{12}$ ceramics remain to be determined. However, it appears that from the CCTO scientific literature, that like in varistors and PTCRs, oxygen may play a key role. For example, in Zang's DSB model the presence of adsorbed oxygen at the grain boundary is crucial.[12] Therefore it was hoped the results of sintering CCTO in flowing oxygen might lead to further insight on the role that oxygen plays in this electrically inhomogeneous system.

However, the results from our research indicate that oxygen plays a role in delaying the onset of abnormal grain growth (AGG). Previous research had shown that CCTO undergoes AGG due to the presence of CuO as CuO-free CCTO powder develops a microstructure of only fine grains.[18,19] Transmission electron microscopy of bimodal CCTO has shown the presence of a thin layer of Cu atoms at the grain boundary, so that some type of Cu-rich grain boundary liquid phase allows for AGG to occur. Ahn et al.[20] in their work on the kinetics of AGG in alumina shows that AGG parallels the mechanisms in a phase transformation where nucleation and growth occurs by showing that alumina undergoes an incubation period followed by substantial growth. However, if the amount of liquid at the grain boundaries is reduced a longer incubation period is needed for AGG to occur but this is followed by AGG creating a microstructure whose overall large grain size is bigger. This is the exact same pattern that is shown in Figure 1. After 2 hours of sintering in air a bimodal-grain microstructure has developed consisting of 25% large grains while the oxygen-sintered material (Figure 1 (a & b)) is still in its

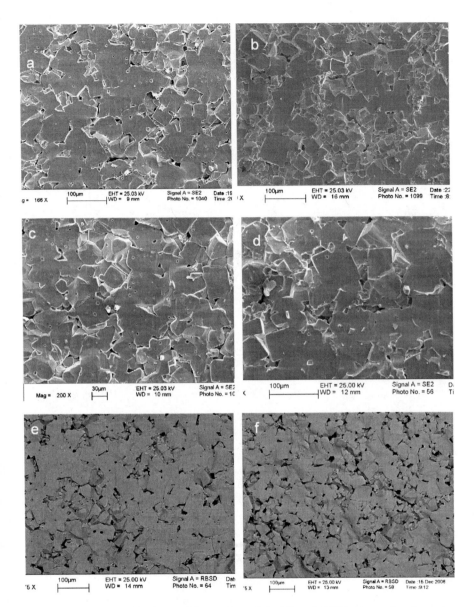

Fig. 5 SEM fractographs of CaCu$_3$Ti$_4$O$_{12}$ sintered for 16h at 1100°C using undoped powders (figs. a, c, e) and doped powders (fig. b- hafnia-doped, fig. d- calcia-doped, fig. f- TiO$_2$-doped).

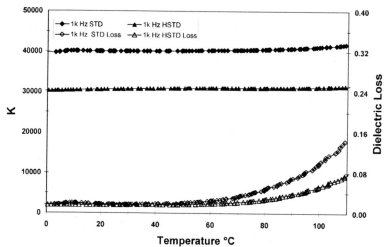

Fig. 6 Temperature dependence of dielectric constant and loss for STD and hafnia-doped HSTD CCTO ceramics sintered at 1100°C for 16 h in air.

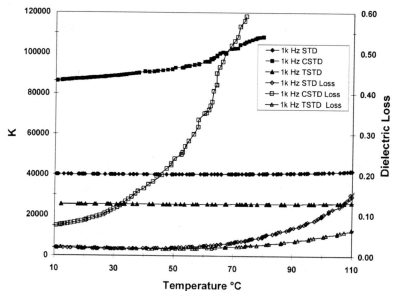

Fig. 7 Temperature dependence of dielectric constant and loss for STD, calcia-doped CSTD, and titania-doped TSTD CCCTO ceramics sintered at 1100°C for 16 h in air.

incubation stage of being 100% fine-grained. After 8 hours the CCTO sintered in air is almost 100% large grains while the oxygen-sintered material has a large grain volume of 35%. It is also interesting to note that the grains are bigger for this material and a lot are more porous indicating coarsening of the grain was outpacing sintering. After 16 hours both materials consist of 100% large grains and as-expected the average grain size was larger for the oxygen-sintered material.

It appears that by sintering in oxygen the amount of Cu-rich liquid phase is being reduced. In air cupric oxide reduces to cuprous oxide at 1026°C.[21] Cuprous and cupric oxide form an eutectic around 1091°C which some variant of the resultant liquid phase going to the grain boundaries to assist in AGG.[22] However, as shown in the phase diagrams, with increasing partial pressure of oxygen the disassociation of cupric to cuprous oxide occurs at higher temperatures which would raise the eutectic temperature also. This could decrease the amount of liquid phase available for AGG and lead to the observed results.

Sintering in oxygen led to small improvements in dielectric loss behavior. As shown in Fig. 2 the dielectric loss is relatively stable and quite low (0.012) which is ideal but at a drastic cost in the reduction of permittivity. In this case, oxygen inhibited AGG and led to dielectric properties similar to those observed for samples sintered in air at 1000°C.[8] In the case of the samples sintered for 8 and 16 hours the dielectric loss exhibits similar behavior where the loss value at room temperature for the oxygen-sintered material is similar to the air-sintered CCTO ceramic but is less sensitive to temperature as shown (see Fig. 3 and 4). This trend was observed by Fang et al.[7] and they showed that insulative boundaries with lower resistivity are more susceptible to an increased temperature dependency on dielectric loss. Therefore it is possible that sintering in oxygen may lead to a more insulative boundary by adsorption of more oxygen at the Schottky barriers or by oxygenation of the semiconducting grains.[12, 23]

The Effect of Dopants on Dielectric Loss, Microstructure, and Permittivity

It is well known that dopants play a key role in the microstructural development and electrical properties of titanates and DSB materials like varistors.[13] Various researchers have explored the role of doping CCTO with Fe, Co, Ni, Nb, Sr, La, V, and Mn.[6,24] However, their interest has been the role they play in affecting dielectric constant or physical properties. Cann et al.[4] found that they could reduce dielectric loss via doping CCTO with zirconia. However, there was close to a 50% reduction in permittivity to 5000 and the material they were working was fine-grained CCTO. Doping with 0.2 w/o hafnia was tried for two reasons. First of all, it is isostructural with zirconia and it is a low loss material (0.007- 1k Hz).[25] The other reason being to see the effect of hafnia on the dielectric properties and microstructure of a higher dielectric large-grained CCTO. No hafnia or hafnia-rich phases were observed in the bulk or grain boundaries. The reduction in grain size from 38 to 30 microns (see Figs.5 (a & b)) implies that some of the hafnia is going into solid solution causing a reduction in grain growth via a solute drag mechanism. The 27% reduction grain size matches well with the 30% reduction in dielectric constant (see Fig. 6). This is expected because in IBLC theory the effective dielectric constant is enhanced by a factor of the grain size divided by the thickness of the insulating boundary layer.[26] Assuming that the boundary layer thickness is the same for both samples the effective dielectric constant should be reduced by the ratio of the grain size of the doped-sample divided by the grain size of the standard sample.

The slightly smaller value of dielectric loss of 0.016 at room temperature of the HSTD CCTO may be due to a reduction in internal porosity of the grains (see Figs a, c, & f vs b). Small changes in porosity can effect dielectric loss.[27] However, doping with hafnia led to a definitive decrease in the sensitivity of dielectric loss with temperature. As shown in Fig. 6 the dielectric loss of the HSTD material is 50% of that of the STD CCTO at 110 °C showing a loss of only 0.08. Cann *et al.*[4] saw a similar trend doping CCTO with zirconia. They observed a dramatic reduction in low frequency loss but not high frequency loss which implies that the resistivity of the grain boundary was augmented. A more insulative boundary phase would lead to a lower temperature dependence on the loss. The mechanism for increasing the boundary resistance is unclear but could be related either to altering the defect chemistry at the boundary layer or just the presence of a low-loss material like hafnia at the boundary layer.

Doping CaCu$_3$Ti$_4$O$_{12}$ with large amounts of CaTiO$_3$ has been shown to substantially reduce its dielectric loss, but again with a dramatic decrease in permittivity from 20,000 to 4500.[28] However, no research has been reported on the doping effect of CCTO doped with CaO or TiO$_2$. CaO is used as a dopant in TiO$_2$ varistors to increase the resistivity of the grain boundaries and therefore made it a material of interest.[29] CCTO doped with 0.3 w/o CaO had a dramatic effect on the dielectric properties of CaCu$_3$Ti$_4$O$_{12}$. The dielectric constant more than doubled (see Fig. 7) while the dielectric loss jumped from 0.018 to 0.087. However, the doubling of the permittivity did not match the 19% increase in grain size observed in the CSTD material (see Fig. 5(c & d)). The mechanism for larger grain growth of the calcia-doped material is unknown but it could be related to a change in defect chemistry. Small amounts of Ca can be incorporated in to the B-sites of AB titanates.[30] Charge compensation occurs by creating oxygen vacancy defects. These defects can effect diffusion properties and lead to faster sintering as evidenced by larger grain growth with fewer pores (compare Figs. 5(a & c) with 5(d)). The creation of defects can also explain the higher loss and permittivity values. Formation of more defects can lead to an increase in conductivity which would lead to a higher dielectric constant.[31] Also the presence of oxygen vacancies could decrease the resistivity of the boundary layer which would lead to higher loss and greater temperature sensitivity of loss which is observed for the HSTD CCTO (see Fig. 7).

Ti plays a key role in the microstructural development and electrical properties of various IBLC and PTCR titanates. Ti has been shown to segregate to the grain boundaries in various strontium or barium titanates designed for use as IBLCs.[32] However, no research has been reported on the presence of excess Ti in CaCu$_3$Ti$_4$O$_{12}$. CCTO doped with 0.25 w/o titania did affect the dielectric properties of CaCu$_3$Ti$_4$O$_{12}$. The permittivity dropped by 37% to 25,300 accompanied by a small increase in loss from 0.018 to 0.020. Doping also lowered the dielectric loss temperature sensitivity as the loss of 0.048 at 100°C of the TSTD material was half that of the STD CCTO (see Fig. 7). Grain size measurements indicated a drop in the average grain size of the STD sample from 38 microns to 35 microns (see Fig. 5(e&f)) with the TSTD material, but the statistics indicated that only at the 90% confidence level that the TSTD effected grain growth. It is unclear what the role that titania-doping plays. The decrease in loss-temperature sensitivity implies that the grain boundary resistance has increased. Perhaps Ti segregates to the grain boundary as it does in SrTiO$_3$ and BaTiO$_3$.[32] This could increase the resistivity of the boundary layer as Fang *et al.*[7] has shown for the increase presence of Cu at the grain boundaries of CCTO. If the presence of extra Ti at the grain boundary led to a more effective wider boundary layer then the permittivity would be expected to drop as observed with the TSTD material. The net result of titania-doping was the fabrication of a CCTO which exceeds X7R specifications with a dielectric constant of 25,340 (1 kHz at room temperature) that varies only by ± 5% from -60 to 125°C with a loss between 0.02 and 0.085 and a variation of permittivity at room temperature of ± 6% in the frequency range from 100 Hz to 100 kHz.

SUMMARY

Sintering in oxygen affected the dielectric loss characteristics and microstructure of $CaCu_3Ti_4O_{12}$ ceramics. Sintering in oxygen delayed the onset of AGG leading to a CCTO with a larger average grain size than its counterpart sintered in air at 1100°C for 16 h. Sintering in oxygen lowered the temperature sensitivity of dielectric loss indicating an increase in the resistivity of the boundary layers. $CaCu_3Ti_4O_{12}$ sintered for 16h in oxygen at 1100°C had a giant dielectric constant of over 40,000 and a dielectric loss only of 0.028 (20°C – 1 kHz).

Doping with calcia, hafnia, and titania also influenced the dielectric loss characteristics and microstructure of CCTO. Calcia-doping affects the defect chemistry of CCTO leading to grain growth. It is postulated that it affects the conductivity of the grains which would explain the doubling of its permittivity to over 80,000 accompanied by an increase in loss. Hafnia-doping leads to a reduction in CCTO grain which is accompanied by a similar reduction in permittivity as predicted by IBLC theory. It also lowered the sensitivity of loss with temperature. Doping with titania led to an even lower loss-temperature sensitivity resulting in a $CaCu_3Ti_4O_{12}$ ceramic with a giant dielectric constant of 25,340 that varies only ± 5% in the X7R range whose dielectric loss at 20°C (1 kHz) is 0.020 and only 0.048 at 100°C. In the future, a better understanding of the nature of the insulative boundaries and defect chemistry of CCTO should yield an engineered-CCTO with lower loss without a reduction in its colossal dielectric constant.

REFERENCES

[1]M.A. Subramanian, D. Li, N. Duan, B.A. Reisner, and A.W. Sleight, "High Dielectric Constant in $ACu_3Ti_4O_{12}$ and $ACu_3Ti_3FeO_{12}$ Phases," *J. of Solid State Chem.*, **151**, 323-25 (2000).

[2]C.C. Homes, T. Vogt, S.M. Shapiro, S. Wakitomo, A.P. Ramirez, "Optical Response of High-Dielectric-Constant Perovskite-Related Oxide," *Science*, **293**, 673-76 (2001).

[3]B.A. Bender and M.-J. Pan, "The Effect of Processing on the Giant Dielectric Properties of $CaCu_3Ti_4O_{12}$," *Mat. Sci. Eng. B.*, **117**, 339-47 (2005).

[4]D.P. Cann, S. Aygun, and X. Tan, "Dielectric Properties of the Distorted $CaCu_3Ti_4O_{12}$," *Extended Abstracts of the 11th US-Japan Seminar on Dielectric and Piezoelectric Ceramics*, 153-56 (2003).

[5]T.-T. Fang, L.-T. Mei, H.-F. Ho, "Effects of Cu Stoichiometry on the Microstructures, Barrier-Layer Structures, Electrical Conduction, Dielectric Responses, and Stability of $CaCu_3Ti_4O_{12}$," *Acta Mat.*, **54**, 2867-75 (2006).

[6]R.K. Grubbs, E.L. Venturini, P.G. Clem, J.J. Richardson, B.A. Tuttle, and G.A. Samara, "Dielectric and Magnetic Properties of Fe- and Nb-doped $CaCu_3Ti_4O_{12}$," *Phys. Rev. B*, **72**, 104111-1-11 (2005).

[7]T.-T. Fang and H.K. Shiau, "Mechanism for Developing the Boundary Barrier Layers of $CaCu_3Ti_4O_{12}$," *J. Am. Ceram. Soc.*, **87**, 2072-79 (2004).

[8]T.B. Adams, D.C. Sinclair, and A.R. West, "Giant Barrier Layer Capacitance Effects in $CaCu_3Ti_4O_{12}$ Ceramics," *Adv. Mater.*, **14**, 1321-23 (2002).

[9]L. He, J.B. Neaton, M.H. Cohen, and D. Vanderbilt, "First-Principles Study of the Structure and Lattice Dielectric Response of $CaCu_3Ti_4O_{12}$," *Phys. Rev. B*, **65**, 214112-1-11 (2002).

[10]S.Y. Chung, I.-D. Kim, and S.J.L. Kang, "Strong Nonlinear Current-Voltage Behaviour in Perovskite-Derivative Calcium Copper Titanate," *Nature Materials*, **3**, 774-78 (2004).

[11]T.B.Adams, D.C. Sinclair, A.R.West, "Characterization of Grain Boundary Impedances in Fine- and Coarse-Grained $CaCu_3Ti_4O_{12}$ Ceramics", *Phys. Rev. B*, **73**, 094124-1-9 (2006).

[12]G. Zang, J. Zhang, P. Zheng, J. Wang, and C. Wang, "Grain Boundary Effect on the Dielectric Properties of $CaCu_3Ti_4O_{12}$ Ceramics," *J. Phys. D. Appl. Phys.*, **38**, 1824-27 (2005).

[13]D.R. Clarke, "Varistor Ceramics," *J. Am. Ceram. Soc.*, **82**, 485-502 (1999).

[14]R.C. Buchanan, *Ceramic Materials for Electronics*, Marcel Dekker, New York, pp. 377-431 (2004).

[15]G.V. Lewis, C.R. Catlow, and R.E. Casselton, "PTCR Effect in $BaTiO_3$," *J. Am. Ceram. Soc.*, **68**, 555-58 (1985).

[16]S. Ananta and N.Thomas, "Relationships Between Sintering Conditions, Microstructure and Dielectric Properties of Lead Iron Niobate," *J. Eur. Ceram. Soc.*, **19**, 1873-81 (1999).

[17]C.C. Chiu, and S.B. Desu, "Microstructure and Properties of Lead Ferroelectric Ceramics ($Pb(Fe_{0.5}Nb_{0.5})O_3$), *Mat. Sci. Eng.*, **B21**, 26-35 (1993).

[18]B.A. Bender and M.-J. Pan, "The Effect of Starting Powders on the Giant Dielectric Properties of the Perovskite $CaCu_3Ti_4O_{12}$," *Ceram. Eng. Sci. Proc.*, **26**, 101-08 (2005).

[19]B.A. Bender and M.-J. Pan, "The Effect of Sintering Conditions and Starting Powders on the Giant Dielectric Properties of $CaCu_3Ti_4O_{12}$," *Extended Abstracts 12th US-Japan Seminar on Dielectric and Piezoelectric Ceramics*, 313-16 (2005).

[20]J.H. Ahn, J.-H. Lee, S.-H. Hong, N.-M. Hwang, and D.-Y. Kim, "Effect of the Liquid-Forming Additive Content on the Kinetics of Abnormal Grain Growth in Alumina," *J. Am. Ceram. Soc.*, **86**, 1421-23 (2003).

[21]E.M. Levin, C.R. Robbins, H.F. McMurdie, and M.K. Reser (ed.), *Phase Diagrams for Ceramists*, The American Ceramic Society, figs. 2069 and 2105 (1968).

[22]F.-H. Lu, F.-X. Fang, Y.-S. Chen, "Eutectic Reaction Between Copper Oxide and Titanium Dioxide," *J. Eur. Ceram. Soc*, **21**, 1093-99 (2001).

[23]D.C. Sinclair, T.B. Adams, F.D. Morrison, and A.R. West, "$CaCu_3Ti_4O_{12}$: One-Step Internal Barrier Layer Capacitor," *Appl. Phys. Lett.*, **80**, 2153-55 (2002).

[24]D. Capsoni, M. Bini, V. Massarotti, G. Chiodelli, M.C. Mozzatic, and C.B. Azzoni, "Role of Doping and CuO Segregation in Improving the Giant Permittivity of $CaCu_3Ti_4O_{12}$," *J. Solid State Chem.*, **177**, 4494-500 (2004).

[25]J.R. Szedon and W.L. Takei, "Dielectric Films for Capacitor Applications in Electronic Technology," *Proc. IEEE*, **59**, 1434-39 (1971).

[26]C.-F. Yang, "Improvement of the Sintering and Dielectric Characteristics of Surface Barrier Layer Capacitors by CuO Addition," *Jpn. J. Appl. Phys.*, **35**, 1806-13 (1996).

[27]D. S. Krueger, R. V. Shende, and S. J. Lombardo, "Effect of Porosity on the Electrical Properties of Y_2O_3-Doped $SrTiO_3$ Internal Boundary Layer Capacitors," *J. Appl. Phys.*, **95**, 4310-15 (2004).

[28]S. Guillemet-Fritsch, T. Lebey, M. Boulos, and B. Durand, "Dielectric Properties of $CaCu_3Ti_4O_{12}$ Based Multiphased Ceramics," *J. Eur. Ceram. Soc.*, **26**, 1245-57 (2006).

[29]M.F. Yen and W.W. Rhodes, "Preparation and Properties of TiO_2 Varistors," *Appl. Phys. Lett.*, **40**, 536-37 (1982).

[30]Y. Li, X. Yao, and L. Zhang, "Dielectric Properties and Microstructure of Magnesium-Doped $Ba_{1+k}(Ti_{1-x}Ca_x)O_{3-x+k}$ Ceramics," *Ceram. Intl.*, **30**, 1283-87 (2004).

[31]W.D. Kingery, H.K. Bowen, and D.R. Uhlmann (ed.), *Introduction to Ceramics*, Wiley & Sons, NY (1976) p. 960.

[32]Y.-M.Chiang and T. Takagi, "Grain-Boundary Chemistry of Barium Titanate and Strontium Titanate: I, High-Temperature Equilibrium Space Charge," *J. Am. Ceram. Soc.*, **73**, 3278-85 (1990).

Electroceramic Materials
for Sensors

MULTIFUNCTIONAL POTENTIOMETRIC GAS SENSOR ARRAY WITH AN INTEGRATED HEATER AND TEMPERATURE SENSORS

Bryan M. Blackburn, Briggs White, and Eric D. Wachsman
UF-DOE High Temperature Electrochemistry Center
Department of Materials Science and Engineering
University of Florida
Gainesville, Florida, USA

ABSTRACT

Solid-state potentiometric gas sensors show much promise for detecting pollutants such as NO_x, CO, and hydrocarbons in ppm level concentrations for exhaust monitoring. The selectivity of these sensors varies with temperature; therefore, sensor arrays can be constructed with integrated heaters to control the temperature of each sensing electrode. The array signals can then be entered into linear algorithms to determine the concentrations of individual species.

The array consisted of two La_2CuO_4 electrodes and a Platinum reference electrode all on the same side of a rectangular, tape-cast YSZ substrate. Platinum resistor elements were used as heaters and/or temperature sensors to control and monitor the temperature of the sensing electrodes. Finite Element Modeling was used to predict temperature profiles within the array. The array was then designed to keep one La_2CuO_4 electrode hot with respect to the other two electrodes. The sensor array configuration avoids complexities that might increase manufacturing costs as found in other designs because all electrodes are in the same gas environment with heater and temperature sensors on the opposite side of the substrate. The results of this work demonstrate that a gas sensor array with sensing electrodes kept at different temperatures can yield a device capable of selectively determining NO and NO_2 concentrations.

INTRODUCTION

Solid-state potentiometric gas sensors base on La_2CuO_4 (LCO) have shown much promise for the monitoring of NO_x levels in simulated combustion exhaust[1]. They are sensitive to ppm levels of NO_x and CO and have fast response times[1,2]. Additionally, they are insensitive to vacillating O_2 and CO_2 concentrations[1,3]. However, the selectivity of these sensors is currently inadequate for commercial application[3]. In fact, poor selectivity hinders most solid-state potentiometric pollutant sensors[4].

One approach for improving selectivity is the use of an array with multiple sensing electrodes, each with a different selectivity. During operation the signals from the array are entered into linear algorithms established during an initial calibration process. The array can then be used to find the concentrations of each measurant (pollutants in this case, e.g., NO). This approach was used for quantitative determination of CO, NO, NO_2, and O_3 in ambient air with resistance-type sensing elements[5,6].

The selectivity of individual sensing elements can be adjusted with temperature to achieve sufficient signal disparity for accurately calculating pollutant concentrations. LCO is sensitive to NO and NO_2, but the sensitivity varies with temperature. At higher temperatures, a loss in NO sensitivity results in improved NO_2 selectivity. To achieve maximum selectivity for each gas, an array can have two La_2CuO_4 electrodes at different temperatures using a heater.

BACKGROUND

Potentiometric gas sensors based on measuring the potential difference between a semiconducting metal oxide and a noble metal pseudo-reference electrode in the same gas environment offer highly selective devices that are easily manufactured and robust in harsh environments[7,8]. These devices were originally thought to be governed solely by mixed potential, where an electromotive force (EMF) is produced at the gas/electrode/solid-electrolyte interface as a result of competing anodic and cathodic reactions occurring at a single electrode[9]. Recent evidence, however, suggests that a more general theory ("Differential Electrode Equilibria") describes sensor responses that cannot be explained with the mixed potential theory alone. The Differential Electrode Equilibria theory predicts that the EMF of an electrode can result from catalytic reactions, semiconductor band-bending, and/or electrocatalytic (mixed potential) reactions[2,10].

Platinum Temperature Sensors and Heating Elements

Platinum was selected for the fabrication of heating elements and temperature sensors. Platinum is an industry standard for high-temperature resistance-temperature-devices (RTD) and as heating elements in gas sensors because of durability and chemical and thermal stability. For the initial design of a sensor array, thermal analysis can be a useful tool to determine the temperature profile of the device when heated locally. The sensing electrodes can then be spaced such that each is at the optimum temperature.

Surface temperature measurements are very difficult and some of the best methods available include use of optical infrared sensors and RTDs. Below approximately 400 °C the resistance of Platinum has a linear dependence on temperature. However, above this temperature, further heat loss causes the linear model to deviate from experimental data, and a better model is

$$R(T) = a\left(1 + bT - cT^2\right)$$ (1)

where a, b, and c are empirical coefficients[10]. After the data is fit to the model, software can calculate the surface temperature during sensor operation using the coefficients from (1) and resistance measurements of the Platinum elements.

Finite Element Modeling (FEM) was performed using the Partial Differential Equation Tool in MatLAB 7.0 from Mathsoft. FEM requires entry of the device geometry (e.g., heater and substrate dimensions), boundary conditions, and material properties. Neumann boundary conditions (i.e., heat flux or (convective) heat transfer) were specified on the boundaries of the YSZ substrate. Dirichlet boundary conditions (i.e., specific temperatures) were entered for the Platinum heater. Material properties used in the analysis are listed in Table I.

Table I. Material properties for 8 mol % YSZ (Marketech International) used in FEM.

Density (g/cm^3)	Thermal Conductivity (W/(m °C))	Heat Capacity (J/g °C)	Length (mm)	Width (mm)	Thickness (mm)
5.85	2.8	0.500	20	12	0.10

An initial analysis predicted that the thermal gradient between the top and bottom of the substrate is negligible because the thickness is so small. This means that temperature measurements with the Platinum elements can provide good estimates for the real temperature of the sensing electrodes, even though they are on opposite sides of the substrate.

Assuming thermal radiation is negligible, heat transfer can occur as conduction through the material or gas. At higher temperatures, convection heat loss through the gas becomes increasingly important[10]. Because the temperature through the thickness is taken as constant, a two-dimensional differential equation for heat transfer can be used to predict the surface temperature profile at a specific heater temperature. The time-dependent differential equation is

$$\frac{\partial T}{\partial t} - \alpha \left[\frac{\partial^2 T}{\partial x^2} + \frac{\partial^2 T}{\partial y^2} \right] + \frac{hAT}{\rho CV} = \frac{hAT_{gas}}{\rho CV} \tag{2}$$

$$\alpha = \frac{\kappa}{\rho C} \tag{3}$$

where h is the heat transfer coefficient of the gas, κ is the material's thermal conductivity, ρ is the density of the material, A is the area of the surface perpendicular to heat flux, and V is the volume of the structure. The heat transfer coefficient depends on testing conditions and was therefore estimated from experimental results.

Heating elements can be used not only to heat another object but also simultaneously as a temperature sensor[11,12]. If the resistance of the heater can accurately be determined (e.g., using a four-wire method), then the temperature of the Platinum element can be calculated. Resistance increases as current is supplied to the heater because of Joule heating. This does not greatly affect the voltage or current measurements. That is to say, the measurements represent the actual current in the circuit and voltage drop across the heater. Therefore the calculated resistance, and hence temperature, of the heater represents the real value.

EXPERIMENTAL

Various Platinum elements were screen-printed in serpentine and C-shaped patterns on planar YSZ substrates and calibrated using custom software. The software also calculated temperatures for each electrode during the ramping of the heater and sensor tests. The array was initially tested with the furnace at 450 °C, 500 °C, and 550 °C to gauge performance with all electrodes at the same temperature. Calculated temperatures agree with results from FEM using the appropriate geometries and material properties. Once repeatability and stability were confirmed, different coplanar sensing electrodes were tested with a heater on the opposite side. As an initial trial, LCO was selected as a NO_x sensing element. The powder was synthesized using a wet-chemical route and characterized in past experiments[4].

The gas sensor array consisted of two LCO sensing electrodes (LCO1 (hot) and LCO2 (cold)) and a Platinum pseudo-reference on the top with C-shaped Platinum elements on the opposite side aligned with the electrodes. Many samples were made with four Platinum elements on the back, but for these experiments only three were used (one heater/temperature sensor and two temperature sensors). Figure 1 shows a top view and cross-section of the array.

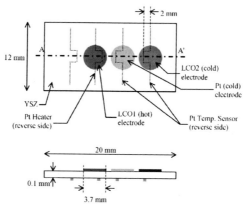

Figure 1. Schematic of gas sensor array showing top view and cross-section A-A'.

Experimental Setup

The experimental setup consisted of a quartz furnace tube with sample holder and wire guides, two computers, and a bank of mass flow controllers with control box. For these initial experiments a Keithley 2400 Sourcemeter, Keithley 2000 Digital Multimeter (DMM) with 10-channel scancard, and a Keithley 2000-20 DMM with 20-channel scancard were used to control the heater, monitor the electrode temperatures, and acquire the OCP measurements from the sensing electrodes. Two DMMs were used to avoid any possible complications associated with differences in the measurement ranges for the Platinum elements and the sensing electrodes. The sensor array was connected to the instruments with a custom wire-harness.

Platinum Element Calibration

Custom software measured the resistances of the Platinum elements as the furnace ramped to 750 °C from room temperature. Data was acquired using a Eurotherm 2408 furnace controller and a Keithley 2000-20. The resistance data was then fit to equation (1) using the Curve Fit Tool in MatLAB 7.0. Other custom software was then used to verify calibration by calculating the temperature of Platinum elements while ramping the furnace to various setpoints between 450 °C and 650 °C. This allowed a direct comparison of the predicted temperatures with the value measured using a thermocouple in the furnace.

Heater Control And Sensor Experiments

Custom software applied voltage incrementally across the Platinum heater aligned with LCO1 until various setpoints were reached. The furnace ambient was maintained at 450°C during all tests. The temperature of each Platinum element was monitored simultaneously during the heater ramp and sensor experiments as described previously. During the heater ramp 3 % O_2 and 97 % N_2 flowed through the quartz tube. At each setpoint, the system was allowed to equilibrate for approximately one hour. Custom software then stepped gas concentrations of NO and NO_2 separately in 3 % O_2 with N_2 balance, while measuring the corresponding open-circuit potential with a Keithley 2000. The total flow rate was always kept equal to 100 sccm.

RESULTS

The curve fits for the Platinum elements were very good ($R^2=1$) and allowed accurate determination of surface temperatures. Figure 2 shows the results of the experiment to verify that the Platinum elements calibration was successful. As evident in the plot, the software calculated the temperature of the Platinum elements very accurately. Deviation from the actual furnace temperature was greatest during the furnace ramp of 5 °C/min and never exceeded ± 3 °C. This deviation was most likely a result of the finite time needed for the Platinum to absorb the heat. Temperatures remained relatively constant during all tests for a given furnace setpoint.

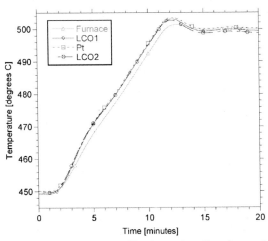

Figure 2. Verification of Platinum element calibration and quality of curve fits to equation (1).

At three voltage setpoints for the heater, the temperatures of the sensing electrodes and the power delivered to the heater were estimated as shown Table II. The temperatures come from the custom software calculations based on resistance measurements of the Platinum elements.

Table II. Surface Temperature Measurements And Heater Power At Three Voltage Setpoints.

Setpoint	Applied Voltage	Heater Power	T_{LCO1}	T_{Pt}	T_{LCO2}
1	1.5 V	70 mW	460°C	452°C	450°C
2	2.5 V	185 mW	484°C	454°C	451°C
3	3.5 V	360 mW	515°C	457°C	452°C

The FEM results agree closely with the measured values (T_{LCO1}, T_{Pt}, and T_{LCO2}) in Table II, which represent average temperatures for the Platinum elements. The temperature profile along the A-A' axis (figure 1) is shown in figure 3. This plot is from FEM with the heater temperature at a uniform 515 °C. The temperature remains fairly constant at the heater and LCO1 electrode, and drops off rapidly as distance from the heater increases.

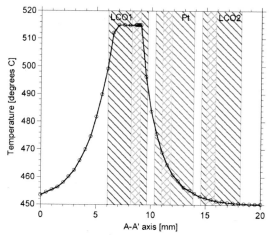

Figure 3. Temperature profile along the A-A' axis from FEM. Also shown are the locations of the sensing electrodes (diagonal, down to right) and Platinum elements (cross-hatch).

Figure 4 shows the steady-state sensitivity plots for LCO1 and LCO2 versus the Platinum reference electrode for NO and NO_2 concentration (0, 50, 100, 200, 400, and 650 ppm) steps. The responses are positive for NO at all concentrations but saturate at higher values. For NO_2 the responses are negative at low concentrations and become positive at higher concentrations. As the heater temperature increases, LCO1-Pt sensitivity to NO steps decreases, while the measured value for NO_2 steps still shows sensitivity. LCO2-Pt has a similar sensitivity to NO_2 as LCO1-Pt, though interestingly shows improved sensitivity to NO as heater temperature increases.

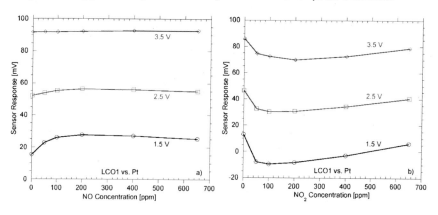

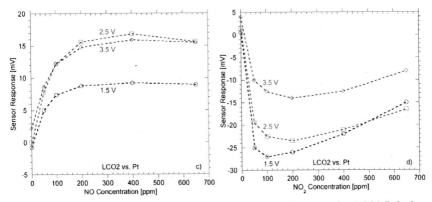

Figure 4. La$_2$CuO$_4$ electrode potential as a function of heater voltage setpoint. LCO1-Pt is shown in a) and b) for NO and NO$_2$, respectively. The response of LCO2-Pt is shown in c) for NO and d) for NO$_2$. See Table II for the corresponding electrode temperatures for each setpoint.

DISCUSSION

The decreasing sensitivity of LCO1-Pt to NO as the heater temperature increases is an improvement over the unheated sensor because the response of LCO versus Platinum is generally positive for increasing concentrations of NO and negative for NO$_2$. Therefore, the overall sensitivity is decreased if both gases are present because the responses tend to cancel each other. The sensor array should not have this problem and the NO and NO$_2$ concentrations should be able to be determined with the measurements from both electrode pairs.

An unheated sensor typically returns to a value near zero after the step gas (e.g., NO or NO$_2$) returns to 0 ppm, and only O$_2$ is flowing[3]. This can be explained with the Nernst equation

$$E = \frac{RT}{4F} \ln\left(\frac{p_{O_2}^{LCO}}{p_{O_2}^{Pt}}\right) = \frac{RT_{LCO}}{4F} \ln\left(p_{O_2}^{LCO}\right) - \frac{RT_{Pt}}{4F} \ln\left(p_{O_2}^{Pt}\right) \tag{4}$$

where, for an unheated sensor, the oxygen concentrations and the temperatures for the LCO and Platinum electrodes are equal and therefore the potential (E) is zero[13]. This is the case for the LCO2 versus Platinum electrodes (figure 4) until the third heater setpoint. However, the sensor array results show a rising baseline voltage (figure 4) for LCO1 versus Platinum as the heater setpoint increases. As the temperature difference between the LCO1 and Platinum electrodes increases from 8 °C to 30 °C, the baseline voltage rises from approximately 15 mV to 50 mV. There is also a change from 50 mV to 90 mV as the temperature difference between the electrodes increases to 58°C. This offset from the expected value of zero is most likely a result of the difference in electrode temperatures.

CONCLUSION

A sensor array with integrated Platinum heater and temperature sensors has been fabricated for small size and low power-consumption. As an initial trial, two La$_2$CuO$_4$ electrodes acted as the NO_x sensing elements. The heater and temperature sensor performance were verified with custom software and FEM. Modulating the temperature of individual sensing electrodes improves selectivity and allows detection of NO and NO_2 when combined in an array. In fact additional sensing materials can be added to the array to detect more species (e.g., CO). The sensor array has reference and sensing electrodes in the same environment, allowing coplanar electrodes and, thus, avoiding complex designs which might increase manufacturing costs.

REFERENCES

[1]E.D. Wachsman, "Selective Potentiometric Detection of NO_x by Differential Electrode Equilibria," Solid State Ionic Devices III, The Electrochem. Soc., E.D. Wachsman, K. Swider-Lyons, M.F. Carolan, F.H. Garzon, M. Liu, and J.R. Stetter, **2002-26**, 215 (2003).

[2]E.D. Wachsman and P. Jayaweera, " Selective Detection of NO_x by Differential Electrode Equilibria", Solid State Ionic Devices II - Ceramic Sensors, Electrochem. Soc., E.D. Wachsman, W. Weppner, E. Traversa, M. Liu, P. Vanysek, and N. Yamazoe, Ed., **2000-32** 298-304 (2001).

[3]B. M. White, F. M. Van Assche, E. Traversa, and E.D. Wachsman, "Electrical Characterization of Semiconducting La_2CuO_4 For Potentiometric Gas Sensor Applications," Solid State Ionic Devices IV, Electrochem. Soc., E.D. Wachsman, V. Birss, F.H. Garzon, R. Mukundan, and E. Traversa, **1**, 109 (2006).

[4]W. Gopel, G. Reinhardt and M. Rosch, "Trends in the Development of Solid State Amperometric and Potentiometric High Temperature Sensors," Solid State Ionics, **136**, 519-531 (2000).

[5]M. C. Carotta, C. Martinelli, L. Crema, C. Malagu, M. Merli, G. Ghiotti and E. Traversa, "Nanostructured Thick-Film Gas Sensors for Atmospheric Pollutant Monitoring: Quantitative Analysis on Field Tests," Sens. Actuators B, 76 [1-3] 336-342 (2001).

[6]M. C. Carotta, G. Martinelli, L. Crema, M. Gallana, M. Merli, G. Ghiotti and E. Traversa, "Array of Thick Film Sensors for Atmospheric Pollutant Monitoring," Sens. Actuators B, **68** (1-3), 1-8 (2000).

[7]E. Di Bartolomeo, M. L. Grilli, N. Antonias, S. Cordiner and E. Traversa, "Testing Planar Gas Sensors Based on Yttria-Stabilized Zirconia With Oxide Electrodes in the Exhaust Gases of a Spark Ignition Engine," Sensor Letters, **3** (1), 22-26 (2005).

[8]M. L. Grilli, E. Di Bartolomeo, A. Lunardi, L. Chevallier, S. Cordiner and E. Traversa, "Planar Non-Nernstian Electrochemical Sensors: Field Test in the Exhaust of a Spark Ignition Engine," Sens. Actuators B, **108** (1-2), 319-325 (2005).

[9]E. Di Bartolomeo, M.L. Grilli, and E. Traversa, "Sensing Mechanism of Potentiometric Gas Sensors Based on Stabilized Zirconia with Oxide Electrodes: Is It Always Mixed Potential?" J. of Electrochem. Society, **151** (5), H133-H139 (2004).

[10]I. Simon, N. Barson, M. Bauer, U. Weimar, "Micromachined Metal Oxide Gas Sensors: Opportunities to Improve Sensor Performance," Sens. and Actuators B, **73**, 1-26 (2001).

[11]M. Baroncini, P. Placidi, G.C. Cardinali, and A. Scorzoni, "Thermal Characterization of a Microheater for Micromachined Gas Sensors," Sens. and Actuators A, **115**, 8-14 (2004).

[12]W. Chung, J. Lim, D. Lee, N. Miura, and N. Yamazoe, "Thermal and Gas-Sensing Properties of Planar-Type Micro Gas Sensor," Sens. and Actuators B, **64**, 118-123 (2000).

[13]N. Rao, C.M. van den Bleek, J. Schoonman, and O.T. Sorensen, "A Novel Temperature-gradient $Na+$-Beta"-Alumina Solid Electrolyte Based SO_x Gas Sensor Without Gaseous Reference Electrode," Solid State Ionics, **53** (6), 30-38 (1992).

PRUSSIAN BLUE NANOPARTICLES ENCAPSULATED WITHIN ORMOSIL FILM

P. C. Pandey, B. Singh
Department of Applied Chemistry
Institute of Technology, Banaras Hindu University
Varanasi, U.P., India

ABSTRACT

Prussian blue (PB) analogues are representatives of the coordination materials and have shown important contributions in the field of Electrocatalysis and molecular magnets. These materials exhibit unique electrochemical and magnetic behaviors depending on their constituents and ratios of transition metal ions in specific geometrical orientation. In-situ generated Prussian blue nanoparticles reported in this communication were prepared by using a mixture of 3-Aminopropyltrimethoxysilane and potassium ferricyanide that was allowed to react with three different moieties; (a) tetrahydrofuran (THF), (b) cyclohexanone, and (c) ferrous sulphate, resulting in the formation of soluble Prussian blue. The whole content is then allowed to trigger the ormosil film formation in the presence of (3,4-epoxy cyclohexyl)ethyltrimethoxysilane, phenyltrimethoxysilane and HCl. An excellent transparent ormosil films encapsulating Prussian blue nanoparticles were formed in each cases. The particles have been examined by atomic force microscopy (AFM) and transmission electron microscopy (TEM) and have shown excellent varying electrochemical properties. The novel finding on these materials on the electrocatalysis of dopamine and hydrogen peroxide based on electrochemical measurements is reported in this communication. The encapsulation of a well known electrocatalyst palladium within the ormosil film has also been made and presented herein.

I. INTRODUCTION

Most of the exiting technology available in world market either incorporates the contribution of photon or electron which essentially differ each other on their inherent characteristics of motion introducing the concept of vacuum and medium-two opposite mode of transmission. Hence, need of such matrix which facilitates the transport of both within same matrix with practical kinetics fall amongst extremely potent requirement of world technology. We provide a little information on this theme in this communication moving around Prussian blue derived from one of the well known redox mediator potassium hexacyanoferrate(III). Prussian blue was discovered by a painter Diesbach in Berlin[1], and therefore it is also called Berlin blue, while attempting to make paint with a red hue. In general, it is a ferri-ferrocynide or ferro-ferricynide type coordination complex of iron (III) and iron (II), and has been given several different chemical names. Commonly and conveniently it is simply called PB[2]. The characteristics of Prussian Blue (PB) have been thoroughly investigated by many research groups [3-6] involved in the development of novel sensing devices. In fact, Prussian blue was found to have a catalytic effect on the reduction of H_2O_2. Moreover, the peculiar cubic geometry of the PB molecules seems to be the cause for an effective electrochemical selectivity. The physical and catalytic properties of the Prussian blue are function of the protocols of its preparation. We report herein the preparation of transparent Prussian blue in-situ generated during organically modified sol-gel glass film assembling.

The recent advances in sol–gel entrapment of biomolecules have been encouraging and have offered a myriad of opportunities. One of such attractions is the development of an electrochemical biosensor that involves the coupling of biological components with polarizable or nonpolarizable electrodes, where the use of sol–gel glasses has received great attention because of their possible applications in commercialization. However, the development of such sol–gel glass-based biosensors is currently restricted mainly due to two major problems: (1) requirement of controlled gelation of the soluble components under ambient conditions, and (2) preparation of a sol–gel glass of smooth surface, controlled thickness and porosity. Additionally, the stability of the biological element within the sol–gel network is another need to develop such sensors on a commercial scale. Apparently the synthesis of a suitable biocompatible sol–gel glass of desired thickness and porosity is of considerable interest. The soluble materials leading to the formation of sol–gel glasses are the derivatives of alkoxysilane. These alkoxysilanes in acidic and sometimes in basic media generate solid-state networks whose physical structure can be compared to that of the conventional glass. The introduction of organic functionalities in alkoxysilanes leads to the formation of organically modified sol–gel glasses (ORMOSIL). The choice of organic functionality could be exploited for controlling the properties of the ormosil film from several angles; (1) to control the water wettability and water contact angle of the film, (2) possibility of anchoring target materials within the matrix, (3) possibility of introducing chemical reactivity during sol–gel processing.

Several organo-functional alkoxysilanes are now commercially available. However, the functional groups like -amino, -epoxy, -glycidoxy, and -phenyl, have already shown fertile reactivity in organic modification of sol-gel glasses. Additionally, the availability of a suitable group within the solid-state network provides an advantage for cross-linking of the sensing element to the solid-state network and enforces the application of the material in opto-electrochemistry. The ormosil films for perfect opto-electrochemical applications should have the following properties; (i) optical transparency, (ii) electrical conductivity, (iii) negligible absorption in a desired wavelength range, (iv) presence of sites for electrocatalytic and chemical interaction, and (v) stability for practical application.

Formation of ormosil films for practical application requires an appropriate proportion of hydrophobic and hydrophilic alkoxysilane modifiers. We have been working on such materials [7-9] derived from 3-aminopropyltrimethoxysilane and 2-(3,4-epoxycyclohexyl) ethyltrimethoxysilane suitable for casting over many metal electrodes surface and also studying such film formation over indium–tin oxide electrodes (ITO) that has potentiality in manipulating opto-electrochemistry, which has bright future in developing powerful microsensors in desired configurations. The use of other alkoxysilanes like phenyltrimethoxysilane may lead better film properties with respect to stability and optical transparency. Similarly the use of 3-glycidoxypropyltrimethoxysilane has shown tremendous potentiality for anchoring palladium within ormosil films enforcing electrocatalytic properties in such materials. Further, sensitization of ormosil films using suitable chemicals, especially organic ion-exchangers like Nafion® or ion-carriers like crown ether may provide valuable information, since such chemicals may lead to affect charge transport and ion-transport within the ormosil films. Accordingly, we developed four systems originally derived from 3-aminopropyltrimethoxysilane, 2-(3, 4-epoxycyclohexyl)ethyltrimethoxysilane and phenyltrimethoxysilane. Potassium ferricyanide/ferrocyanide was also encapsulated together with these chemicals for evaluating the performances of such ormosil-modified electrodes electrochemically and subsequently studied the oxidation of hydrogen peroxide[10-11]. The results reported in earlier publications [10-11] directed our attention to investigate the followings; (1) effect of variation in potassium ferricyanide concentration within ormosil films, (2) effect of the variation in dibenzo-18-crown-6 concentration within ormosil films, (3) effect of pH on the electrochemical oxidation of hydrogen peroxide at the surface of these modified electrodes, and (4) the variation in the nano-/micro-structure of the films in the presence of these dopants, if any. Since both an electron-transfer relay and a chemical sensitizer (Nafion® /crown ether) were encapsulated together within an ormosil film, it was also intended to understand the electrochemical performances of such modified electrodes when these two materials were separated within well-defined domains with the possibility of interaction of the electrochemical reactions. Accordingly a concept of bilayer structure of ormosil film was introduced. It was also reported that the microstructure of ormosil films tend to alter with time due to conversion into xerogel. It was further intended to understand the variation in microstructure of ormosil films under varying experimental conditions based on atomic force microscopy. Some of the findings on these lines are reported in this communication.

II. EXPERIMENTAL

Material:

3-Aminopropyltrimethoxysilane, dibenzo-18-crown-6 and phenyltrimethoxysilane were obtained from Aldrich Chemical Co. 2-(3,4-Epoxycyclohexyl)ethyltrimethoxysilane was obtained from Fluka. The aqueous solutions were prepared in triply-distilled water. All other chemicals employed were of analytical grade. A Julabo water thermostat was used to control the temperature of the reaction cell during measurements.

Construction of Ormosil Modified Electrodes:

The protocols for the construction of these four types of ormosil-modified electrodes are as follows; System -1: 3-aminoproyltriethoxy silane (70 µl), 2-(3,4-Epoxycyclohexyl)ethyl trimethoxysilane (10 µl) and , phenyltrimethoxy silane (5 µl), distilled water (240 µl), 10 mM aqueous solution of $K_3Fe(CN)_6$ (60 µl) and HCl (5 µl); System -2; 3-aminoproyltriethoxy silane (70 µl), 2-(3,4-Epoxycyclohexyl)ethyl trimethoxysilane (10 µl) and phenyltrimethoxy silane (5 µl), distilled water (235 µl), 10 mM aqueous solution of $K_3Fe(CN)_6$ (60 µl), ethanolic solution of Nafion® (5 µl) and HCl (5 µl); System-3: 3-aminoproyltriethoxy silane (70 µl), 2-(3,4-Epoxycyclohexyl)ethyltrimethoxy silane (10 µl) and , 2 mg dibenzo-18-crown-6 dissolved in 200µl phenyltrimethoxy silane (5µl), distilled water (240 µl), 10 mM aqueous solution of $K_3Fe(CN)_6$ (60 µl) and HCl (5 µl); System -4: THF (10 µl) or cyclohaxanone (5 µl), 3-aminoproyltriethoxy silane (70 µl), 2-(3,4-Epoxycyclohexyl)ethyl trimethoxysilane (10 µl) and , phenyltrimethoxy silane (5 µl), distilled water (240 µl), 10 mM aqueous solution of $K_3Fe(CN)_6$ (60 µl) and HCl (5 µl) were homogenized and 10 µl of the each homogeneous solution was casted over ITO electrodes. The ormosil's film was allowed to dry for 7–10 hours at room temperature and the electrochemistry of encapsulated potassium ferricyanide in each case was examined by cyclic voltammetry.

Single Layer Configuration:

The protocol for the construction of ormosil-modified electrodes were as follows; 3-aminopropyltrimethoxysilane (70 µl), 2-(3,4-epoxycyclohexyl)ethyltrimethoxysilane (10 µl) and phenyltrimethoxysilane (5 µl) containing appropriate concentrations of dibenzo-18-crown-6 [0.4 mg/ml; 1 mg/ml], distilled water (240 µl), 60 µl aqueous solution of $K_3Fe(CN)_6$ of desired concentrations, and HCl (5 µl) were homogenized, and 10 µl of each homogeneous solution was cast over ITO electrodes. The ormosil films obtained were allowed to dry for 7–10 hours at room temperature and the electrochemistry of encapsulated potassium ferricyanide was examined by cyclic voltammetry in each case.

Bilayer Configuration:

Two protocols for bilayer film formation were adopted as follows. Protocol-I: the first layer of ormosil (for all four systems) was made as described above. After drying the films for 7–10 hours, the second layer of ormosil was cast using 5 µl of homogenized ormosil precursors as given in Table I, [i.e. 3-aminopropyltrimethoxysilane (70 µl), 2-(3,4-epoxycyclohexyl)ethyltrimethoxysilane (10 µl), phenyltrimethoxysilane (5 µl), distilled water (300 µl), 60 µl aqueous solution of $K_3Fe(CN)_6$ of desired concentrations, and HCl (5 µl)]. Protocol-II: the first layer of ormosil was made using 5 µl of homogenized ormosil precursors as given in Table I. After drying the films of basically two different types, i.e. having potassium ferricyanide or in-situ generated Prussian blue, for 7–10 hours, the second layer of ormosil was cast using the same amount of homogenized ormosil precursors containing either Nafion[®] or dibenzo-18-crown-6 (Table I).

Table I- Composition of ormosil precursors for making bilayer based systems

Protocol	System	A[*] (µl)	B[*] (µl)	C[*] (µl)	D[*] (µl)	E[*] (µl)	F[*] (µl)	G[*] (µl)	H[*] (µl)	I[*] (µl)
I	1st layer	70	10	5	–	–	60	240	5	–
	2nd layer	70	10	5	–	–	–	300	5	–
I	1st layer	70	10	5	–	5	60	235	5	–
	2nd layer	70	10	5	–	–	–	300	5	–
I	1st layer	70	10	–	5	–	60	240	5	–
	2nd layer	70	10	5	–	–	–	300	5	–
I	1st layer	70	10	5	–	–	60	240	5	5
	2nd layer	70	10	5	–	–	–	300	5	–
II [A]	1st layer	70	10	5	–	–	60	240	5	–
	2nd layer	70	10	5	–	5	–	295	5	–
II [B]	1st layer	70	10	5	–	–	60	240	5	–
	2nd layer	70	10	5	5	–	–	195	5	–
II [C]	1st layer	70	10	5	–	–	60	240	5	5
	2nd layer	70	10	5	5	–	–	295	5	–

A[*], 3-aminopropyltrimethoxysilane; B[*], 2-(3,4-epoxycyclohexyl)ethyltrimethoxysilane; C[*], phenyltrimethoxysilane; D[*], dibenzo-18-crown-6 (2 mg/ml) in phenyltrimethoxysilane; E[*], ethanolic solution of Nafion[®] as supplied by Aldrich; F[*], aqueous solution of potassium ferricyanide (10 mM); G[*], distilled water and H[*], HCl (0.5N), I[*], THF.

For dopamine sensing the modified electrodes were fabricated by using the protocol given in Table II.

Table II- Composition of ormosil precursors for developing dopamine sensor

System	A*	B*	C*	D*	E*	F*	G*	H*
1	70 μl	10 μl	5 μl	--	--	60 μl	240 μl	5 μl
2	70 μl	10 μl	5 μl	--	5 μl	60 μl	235 μl	5 μl
3	70 μl	10 μl		5 μl		60 μl	240 μl	5 μl

A*=3-Aminoproyltriethoxy silane; B*=2-(3,4-Epoxycyclohexyl)ethyl trimethoxysilane; C*= phenyltrimethoxy silane; D*= Dibenzo-18-Crown-6 (2 mg /ml) in phenyltrimethoxysilane; E*= Ethanolic solution of Nafion as supplied by Aldrich; F*= aqueous solution of potassium ferricyanide (10 mM); G*=Distilled water and H*=HCl (0.5 N)

Electrochemical Measurements:
    The electrochemical measurements were performed with a CH Instruments Electrochemical Workstation 660B. A one-compartment cell with a working volume of 3 ml and the ormosil-modifed electrode with an exposed surface area of 2 mm as a working electrode, a Ag/AgCl reference electrode and a platinum foil as an auxiliary electrode were used for the measurements. The cyclic voltammetry was studied between −0.3 and + 0.6 V versus Ag/AgCl. The amperometric measurements using the modified electrode were obtained at a desired operating potential versus Ag/AgCl in the presence and absence of peroxide. The experiments were performed in a phosphate buffer (0.1 M, pH 7.0) at 25 °C. Before each set of measurements the working solution was degassed by purging nitrogen for 15 min.

AFM characterization:
    The topography of the sol–gel surfaces doped with chemical sensitizers was analyzed using a Solver-Pro AFM (NT-MDT, Russia) in the non-contact mode. All scans were taken at room temperature in air using silicon nitride tips (spring constant 40 N/m).

III. RESULTS AND DISSCUSSION
    We have made investigations of the electrochemistry of electron transfer mediators encapsulated within ormosil film derived modified electrodes and based on electrochemical measurements concluded the following: (1) incorporation of palladium within the nanoporous geometry of ormosil facilitates the charge transport with excellent redox electrochemistry of encapsulated mediator by introducing electrocatalytic power in ormosil-modified electrode, (2) incorporation of ruthenium derivatives also facilitates the charge transport within solid-state matrix as well as influences the electrocatalytic efficiency of the material. All these observations reveal the contribution of metal-ceramic composite to improve the electrochemistry and electrocatalytic properties of the modified electrodes. On the other hand, introduction of organic moieties having ion-exchange and ion-recognition sites within ormosil matrix might alter the electrochemistry of ormosil-encapsulated mediators. Accordingly, it was aimed to investigate on these lines in the presence of Nafion® and dibenzo-18-crown-6. We have observed that the ormosil films derived from 3-aminopropyltrimethoxysilane, 2-(3, 4-epoxycyclohexyl)ethyltrimethoxysilane and phenyltrimethoxy silane in acidic medium retain the inherent property of these organic moieties and the resulting films are highly stable for practical applications. Potassium ferricyanide is also encapsulated together with such moieties which generates excellent ormosil film for electroanalytical applications. For easy presentation we

designated the different preparations as systems 1-3. The ormosil film made with $K_3Fe(CN)_6$ only, represents system-1, the same with $K_3Fe(CN)_6$ and Nafion® is system-2, and the film with $K_3Fe(CN)_6$ and dibenzo-18-crown-6 represents system-3. While proceeding ahead on the preparation of ormosil film using dibenzo-18-crown-6, we noticed the problem of its solubility. Fortunately dibenzo-18-crown-6 was a little soluble into phenyltrimethoxysilane and the initial observations were satisfactory with system-3. However, we further noticed that the data on different concentrations of crown ether would be valuable to many electroanalytical applications; accordingly we attempted to increase the concentration of dibenzo-18-crown-6 in ormosil film using tetrahydrofuran as solvent. During such experimentation we observed that potassium ferricyanide get converted into Prussian blue in the presence of 3-aminopropyltrimethoxysilane and tetrahydrofuran or cyclohexanone. Such conversion again directed our attention to study the system derived from only Prussian blue in the ormosil film and further in the presence of varying concentrations of dibenzo-18-crown-6 together with the Prussian blue within the film. It should be noted that during such addition care was always taken to retain the inherent property of dibenzo-18-crown-6 within the film.

Similarly, the redox electrochemistry of potassium hexacyanoferrate-encapsulated modified electrodes was quite interesting and provided valuable information useful for chemical/biochemical sensing. The differential electrochemical sensing of hydrogen peroxide was recorded on these modified electrodes basically due to the participation of Nafion®/crown ether associated with either facilitated diffusion of peroxide or introduction of electrocatalytic property in the electrode materials [9-11]. In order to understand precisely the contributions of each component it was required to control the configuration of the ormosil film, thus we coupled the electron-transfer component and ion-exchange/ion-carrier site in a sequential pattern. Such configuration could easily be availed in a bilayer configuration of the ormosil film formation, thus allowing the sequential interaction events during electrochemical sensing. The single layer and bilayer-based configuration could be schematically represented as follows:

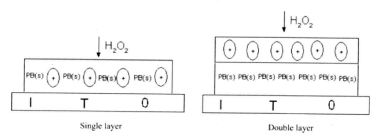

Single layer   Double layer

As a testing ground we first intended to study the performance of the modified electrodes in a bilayer configuration having the first layer with a similar pattern to that reported in earlier publication[9-11], whereas the second layer made by using the composition of a plane ormosil layer that neither contained the electron-transfer component nor the Nafion® or crown ether. Accordingly, protocol-1 as given in Table I was adopted to understand the effect of introduction of bilayer structure on the peroxide sensing. The electrochemical performances of the bilayer modified electrodes were encouraging with following remarkable observations; (i) the performances of the electrodes on electrocatalytic oxidation of hydrogen peroxide was analogous to that of the single layer-based system with a significant decrease in electrocatalytic reduction of hydrogen peroxide, (ii) occurrence of well-defined diffusion-limited condition with a significant increase in the linear range of peroxide sensing, (iii) peroxide sensing was found more selective in the presence of interfering analyte like ascorbic acid, since the double layer caused significant reduction in electrochemical oxidation of ascorbic acid as compared to that of on single layer modified electrodes under similar conditions, and (iv) the kinetics of peroxide reduction was significantly reduced as compared to that on single layer modified electrodes under similar conditions. Such remarkable findings again directed our attention for further investigation on the modified electrodes with a bilayer configuration made by following protocol-II as given in Table I. The origin of protocol-II was as follows. The four systems reported in earlier publication[9-11] were mainly based on the contribution of two important components, an electron transfer relay (potassium ferricyanide) and a chemical sensitizer (Nafion[®]/crown ether). While dealing with another system, potassium ferricyanide was in situ converted into Prussian blue, thus introducing electrocatalytic sites in the electrode material. The results based on cyclic voltammetry are discussed in the following sections on these systems.

Construction of in-situ generated Prussian blue:

Chemistry of electrode-4, prepared by using the reaction product of THF/cyclohexanone, 3-aminopropyltrimethoxysilane and potassium ferricyanide, was found very interesting. Potassium ferricyanide in the presence of 3-aminopropyltrimethoxysilane and either Tetrahydrofuran or Cyclohexanone undergo Prussian blue formation and the yellow colour of potassium ferricyanide changes depending on the nature of electron acceptor participating in Prussian blue formation as shown in Fig. 1a. THF generates light blue color whereas cyclohexanone results dark green color. Fig.1b (I) shows the variation of UV-visible absorption maxima under three different conditions. Curve-1 in Fig.1b (I) and (II) shows the absorption as a function of wavelength of the solution containing potassium ferricyanide and 3-aminopropyltrimethoxysilane before the addition of cyclohexanone/ THF whereas curves 2, 3, and 4 [Fig.1b (II)] show the absorption after the addition of cyclohexanone at the interval of 5 min in each case. Curve 2 in (II), after adding tetrahydrofuran, 3 in (II) shows recording of same system at the interval of 5 min.

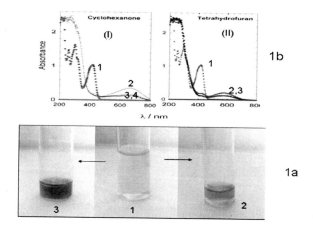

Figure 1. 1(a) Colour change in the formation of charge transfer complex with (1) potassium ferricyanide, (2) tetrahydrofuran and (3) cyclohexanone; 1(b) Absorption spectra of Prussian blue formation in the presence of cyclohexanone (I) and tetrahydrofuran (II).

The variations in the absorption spectra shown in Fig.1b justify the following; (1) the absorption maxima recorded for potassium ferricyanide at 420 nm as curve-1 in both Fig.1b (I) and (II) disappears after the addition of cyclohexanone/THF; (2) The reaction of cyclohexanone to form Prussian blue is faster than that of THF; (3) the reaction product with cyclohexanone causes reappearance of a new peak around 675 nm [Fig.2b (I) curve 2 and 3] whereas THF generates a new peak at 570 nm; (4) The absorption peaks at both 570 nm and 675 nm get disappeared kinetically; (4) variation in absorption in UV region is also seen in the reaction product of cyclohexanone and tetrahydrofuran. Such variation in the kinetics of cyclohexanone interaction reflects charge transfer complexation during Prussian blue formation. Such conclusion is also supported by another spectrophotometric observation of ormosils without and with Prussian blue formation. Fig.1c shows the absorption spectra of ormosil [controle/system-1 (curve-1) and system-4 of cyclohexanone (curev-2) and tetrahydrofuran (curev-3). Disappearance of absorption at 420 nm (curve-1 as observed system-1and appearance of large plateau in UV region in system-4 in both cases (cyclohexanone and tetrahydrofuran) justifies the charge-transfer complexation in Prussian blue formation.

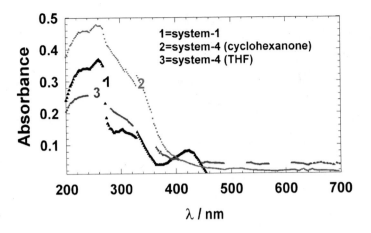

Figure 1(c). Absorption spectra of ormosils

Atomic Force Microscopy of Prussian blue encapsulated ormosil:

It is further important to understand the practical usability of the electrode systems. We have studied the electrochemical performances of the ormosil-modified electrodes as a function of drying time and concluded that it dramatically affected the electrochemical performance. The drying time of 8–10 hours under the laboratory conditions was found suitable of attaining optimum electrochemical performances of the modified electrodes. Reduced drying time caused the leaching out of the electron-transfer mediators from the film, whereas a considerable increase in drying time caused a gradual decrease in electrochemical signal of the ormosil-encapsulated mediators. Such variation is expected to be due to the variation in microstructure of the ormosil film, which is resulted due to the conversion into the xerogel state. However, the Prussian blue developed by other author, as reviewed by Ricci and Palleschi[12], and the material made in the present investigation certainly differ in structural arrangements and properties. Further, the choice of electron donors during Prussian blue formation dramatically alters the visual variation during such material formation, as reported earlier[12]. The use of THF and cyclohexanone showed significant changes in visible absorption spectra during early few minutes. Accordingly, it is remarkable to investigate the variation in microstructure of the films through atomic force microscopy (AFM) on the use of (a) THF, (b) cyclohexanone, (c) crown ether-encapsulated film made with cyclohexanone. These microstructures are shown in Fig. 2a–c respectively.

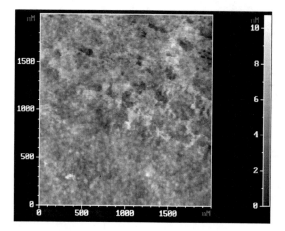

Figure 2a. AFM image of the ormosil film made in a single layer configuration in the absence of crown ether and derived from THF.

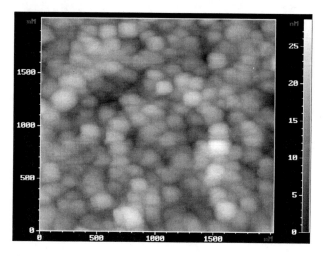

Figure 2b. AFM image of the ormosil film made in a single layer configuration in the absence of crown ether and derived from cyclohexanone

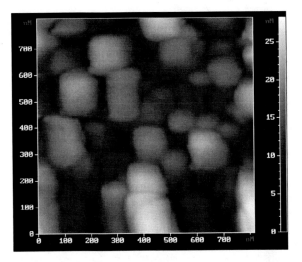

Figure 2c. AFM image of the ormosil film made in a single layer configuration in the presence of crown ether and derived from cyclohexanone

The AFM results have revealed that there was remarkable variation in microstructure when THF was changed by cyclohexanone. Similarly, the introduction of crown ether also significantly altered the microstructure of the film. We are further investigating the reasons of such variation through examination of exact change in crystal structure by transmission electron microscopy (TEM). Also, various other routes have been followed to make new analogues of Prussian blue. The replacement of THF/cyclohexanone by FeSO4 has generated a novel material under the present experimental protocol. The presence of amino-substituted alkoxysilane caused the production of nano-sized Prussian blue analogues derived from ferricyanide or ferrocyanide system on drop-wise addition of FeSO4 or ferric chloride respectively. Further, the use of 3-glycidoxypropyltrimethoxysilane has facilitated the introduction of palladium to enforce the electrocatalysis in hydrogen peroxide oxidation.

Transmission Electron Microscopy:
    The chemistry of in-situ generated Prussian blue formation seems to be very interesting and is the function of component triggering the charge transfer complexation during iron (II) and iron (III) stabilization together with the change in the physical morphology of the material. The present invention incorporates the coupling of 3-aminopropyltrimethoxysilane, potassium ferricyanide and the component namely: (a) THF, (b) cyclohexanone, (c) ferrous sulphate, which triggers the initiation of charge transfer complexation. These components drastically alter the microstructure of the material. A typical example showing the diffraction pattern and microstructure of the material is shown in the Figure 3. In this case the formation of Prussian blue was triggered by using THF. It was found that the size of the material ranges between 50-nm -200 nm.

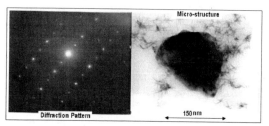

Figure 3. TEM image of ormosil film encapsulated with Prussian blue derived from the use of tetrahydrofuran.

Electrocatalytic sensing of hydrogen peroxide:

Figure 4(a) shows the electrochemistry of potassium ferricyanide encapsulated within these ormosil-modified electrodes (system-1 to system-4) as discussed in the preceding sections. Amongst all these modified electrodes, presence of Nafion[®] within the ormosil film caused better redox electrochemistry of ormosil encapsulated potassium ferricyanide. The peak separation is virtually independent of scan rate which is attributed to ion-exchange behaviour of Nafion[®], whereas the crown ether modified ormosil shows relatively larger peak current and the peak separation changes significantly on scan rate. Relatively greater peak separation in system-3 may be due to the existence of ion-pair and charge separation is possibly introduced within film through the recognition of potassium ion by dibenzo-18-crown-6 whereas ferricyanide ions lie out side of the macrocyclic cavity. This generates an increase in capacitive current as supported by increase in capacitive components in system-3. Such remarkable findings again directed our attention to further investigate the modified electrodes with a bilayer configuration made following protocol-II as given in Table I. The origin of protocol-II was as follows. The four systems reported in earlier publication were mainly based on the contribution of two important components, an electron transfer relay (potassium ferricyanide) and a chemical sensitizer (Nafion[®]/crown ether). While dealing with another system, potassium ferricyanide was in situ converted into Prussian blue, thus introducing electrocatalytic sites in the electrode material.

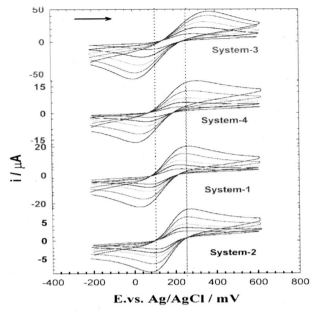

Figure 4(a) Cyclic voltammograms of potassium ferricyanide encapsulated ormosil-modified electrodes prepared under four different conditions (Table I) in phosphate buffer (0.1 M, pH 7.0) at the scan rate of 5, 10, 20, 50 and 100 mV/s.

It was further found that the formation of Prussian blue also caused the electrocatalytic oxidation of ascorbic acid. The result based on the single layer Prussian blue modified electrode is shown in figure 4b in the presence of ascorbic acid (curves 1 and 2) and hydrogen peroxide (curves 1 and 3). The results clearly justify the introduction of electrocatalysis to the ascorbic acid oxidation, although the peroxide sensing was also enforced as compared to that with potassium ferricyanide.

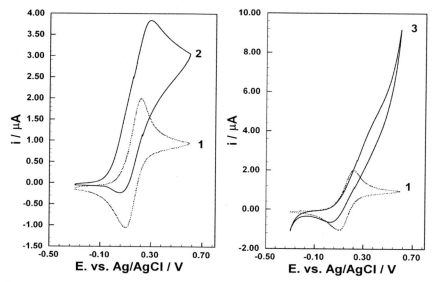

Figure 4(b). Cyclic voltammograms of single layer Prussian blue modified electrode in phosphate buffer (0.1 M, pH 4.0). 1-in absence of hydrogen peroxide/ ascorbic acid, 2-in presence of 1 mM ascorbic acid, and 3-in presence of 1 mM hydrogen peroxide.

Electrocatalytic sensing of dopamine:

The analysis of dopamine is normally corrupted due to the presence of other electro active species especially ascorbic acid present in physiological fluid. However, we investigated the dopamine sensing in the presence of ascorbic acid on a potassium ferricynide and dibenzo-18-crown-6/ Nafion® encapsulated ormosil modified electrode in single layer configuration and the cyclic voltammogrames are presented in Figure 5.

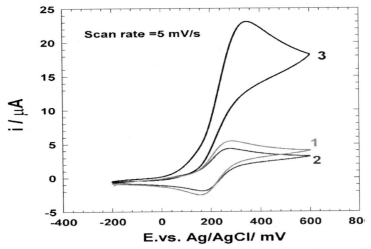

Figure 5. Cyclic voltammograms of system-5 before (curve-1) and after the addition of 0.2 mM ascorbic acid (curve-2) followed by addition of 0.2 mM dopamine (curve 3) in phosphate buffer (0.1 M, pH 7.0) at the scan rate of 5 mV/s.

Curve-1 shows the result before the addition of ascorbic acid whereas curve-2 shows the same after the addition of 1 mM ascorbic acid, which is around three fold higher than the normal physiological ascorbic acid concentration. After the addition of ascorbic acid, instead of increase in amperometric response, the anodic current gets decreased (curve-2; Fig.5). When dopamine (0.2 mM) is added in the system after ascorbic acid addition, there is enough response on dopamine sensing which justifies that the system is not corrupted by the presence of ascorbic acid. However, when the ormosil-modified electrode without dibenzo-18-crown-6 was used for dopamine sensing, the electrode responded to ascorbic acid. Therefore, the selectivity of dopamine sensing justified the participation of dibeno-18-crown-6. Insensitivity of system containing dibeno-18-crown-6 towards the presence of ascorbic acid can be explained as follows. The presence of anion within the ormosil matrix restricts the diffusion of ascorbate anion within the ormosil. Accordingly, reverse in the diffusion of ascorbic acid kinetics as compared to dopamine diffusion kinetics prevail which justify the importance of present system for dopamine sensing. We also observed that increase in crown ether concentration caused relative decrease in ascorbic acid sensing that again justify the contribution of dibenzo-18-crown-6 in dopamine sensing.System-2 also not responded to the ascorbic acid, which is in accordance with the property of Nafion® restricting the diffusion of ascorbic acid within the film.

IV. CONCLUSION

The present paper demonstrates the methods for developing organically-modified sol-gel glass (ORMOSIL) film modified electrodes suitable for electrochemical sensing. As the examples electrochemical sensing of dopamine and hydrogen peroxide is reported. During such fabrication we found in-situ conversion of potassium ferricyanide into Prussian blue like structure that introduced electrocatalysis during electrochemical sensing. The results based on

UV-VIS spectroscopy, Atomic force microscopy and transmission electron microscopy are reported on few systems. These observations suggest the possibility of further enhancing the electrocatalytic efficiency by introducing some known nanomaterials like palladium, gold nanoparticles and carbon nanotubes. Another possibility has also been introduced on incorporating ion recognition sites within ormosil matrix. Some important findings on these lines are reported in this communication.

Acknowledgement
        The authors are thankful to the department of Science & Technology, Govt. of India for financial support

REFERENCES

[1] http://painting.about.com/cs/colourtheory/a/prussianblue.htm

[2] K. R. Dunbar and R. A. Heintz, "Chemistry of Transition Metal Cyanide Compounds: Modern Perspectives", *Prog. Inorganic Chem.*, **45**, 283-391 (1997).

[3] A. A Karyakin, E. E. Karyakina and L. Gorton "Prussian-Blue-based amperometric biosensors in flow-injection analysis" *Talanta*, **43**, 1597 (1996).

[4] A. A. Karyakin, O. V. Gitelmacher and E. E. Karyakina, "Prussian Blue-Based First-Generation Biosensor. A Sensitive Amperometric Electrode for Glucose" *Anal. Chem.*, **67**, 2419 (1995).

[5] K. Itaya, H. Akahoshi and V. D. Neff., "Electrochromism in the mixed-valence hexacyanides. 1. Voltammetric and spectral studies of the oxidation and reduction of thin films of Prussian blue", *J. Phys. Chem.*, **85**, 1225 (1981),

[6] S. A. Jaffari and. A. P. F. Turner, "Novel hexacyanoferrate(III) modified graphite disc electrodes and their application in enzyme electrodes—Part I", *Biosens. Bioelectron.*, **12**, 1 (1997).

[7] P. C. Pandey, S. Upadhyay, I. Tiwari and V. S. Tripathi, "An ormosil based ethanol biosensor", *Anal. Biochem.*, **288**, 39 (2001).

[8] P. C. Pandey, S. Upadhyay, I. Tiwari, G. Singh and V. S. Tripathi, "A novel ferrocene encapsulated palladium-linked ormosil-based electrocatalytic dopamine biosensor", *Sens. Actuators B*, **75**, 48 (2001).

[9] P. C. Pandey, S. Upadhyay, N. K. Shukla and S. Sharma, "Studies on the elecrochemical performance of glucose biosensor based on ferrocene encapsulated ormosil and glucose oxidase modified graphite paste electrode", *Biosens. Bioelectron.*, **18**, 1257 (2003).

[10] P. C. Pandey, B. C. Upadhyay and A .K. Upadhyay, "Differential selectivity in electrochemical oxidation of ascorbic acid and hydrogen peroxide at the surface of functionalized ormosil-modified electrodes", *Anal. Chim. Acta*, **523**, 219 (2004).

[11] P. C. Pandey and Upadhyay B.C., "Studies on differential sensing of dopamine at the surface of chemically sensitized ormosil-modified electrodes", *Talanta*, **67**, 997(2005).

[12] F. Ricci and G. Palleschi, "Sensor and biosensor preparation, optimisation and applications of Prussian Blue modified electrodes", *Biosensor. Bioelectron.*, **21**, 389.

GAS-SENSING PROPERTY OF HIGHLY SELECTIVE NO$_X$ DECOMPOSITION ELECTROCHEMICAL REACTOR

Koichi HAMAMOTO, Yoshinobu FUJISHIRO, and Masanobu AWANO
Advanced Manufacturing Research Institute, National Institute of Advanced Industrial Science and Technology (AIST),
2266-98 Anagahora, Shimoshidami, Moriyama-ku
Nagoya, Aichi 463-8687 JAPAN

ABSTRACT

Yttria-stabilized zirconia (YSZ) electrochemical cell with the self-assembled catalytic layer sensing electrode was fabricated and examined for the mixed-potential-type and the amperometric-type NO$_X$ sensor. This cell shows a high NO sensitivity. The magnitude of EMF response was about -43 mV upon exposure to 100 ppm of NO at 400 °C. The amperometric-type gas sensor shows a high NO sensitivity of 33nA/ppm at especially less than 20ppm of the NO concentration. And, the current response and recovery times were found to be about 0.25 and 0.7 sec, respectively. The availability of the self-assembled catalytic layer electrode for the multiple sensing of NO$_X$ was confirmed. This cell opens up the possibilities for development of highly sensitive NO$_X$ gas sensors.

INTRODUCTION

Nitrogen oxides (NO$_X$) gases have a deleterious impact on the environment as well as human and animal health. Lean-burn gasoline and direct-injection diesel engines have been developed to improve fuel efficiency as well as to reduce CO$_2$ emission form engine, but their development is prevented by their high NO$_X$ emissions. In these engine systems, recently developed NO$_X$ adsorber-catalyst should be used in order to compensate for the low NO$_X$ removal ability of the conventional three-way catalyst under the lean-burn (oxygen rich) condition. In a complete control of this catalyst system, a reliable and responsive NO$_X$ sensor for the on-board diagnostic (OBD) system is required to detect the leakage of NO$_X$ from the NO$_X$ adsorber-catalyst in lean exhaust and signal the need for catalyst regeneration, such as its selective catalytic reduction with HC (1-4) or NH$_3$ (1,5-6). The OBD system is a closed-loop system, mounted on each automobile, to continuously monitor the pollutant concentrations in the exhausts. At present, the solid-state NO$_X$ sensor normally uses a two-chamber O$_2$-pumping potentiometric gas sensor (7-11) for OBD system, which is very expensive in mass production. Meeting the stringent requirements of the automobile industry in terms of manufacturing cost and sensitivity, cheap and innovative gas sensing devices need to be developed.

Recently, we have developed a new type of electrochemical cell with a functional multilayer cathode and achieved drastic improvements of selective NO_X reduction in the presence of excess O_2 (12-16). This reactor can be represented by the following cell arrangement: YSZ | Pt | NiO-YSZ (Catalytic layer) | YSZ | Pt +YSZ. The multilayer cathode includes a NO_X selective catalytic layer whose structure was designed to generate self-assembled nano-pores and nano-Ni particles in the vicinity of NiO-YSZ interfacial regions by the cell operation. The catalytic layer provides a way to suppress the unwanted reaction of O_2 gas decomposition, resulting in increasing the efficiency of NO decomposition. The cell showed that the decomposition selectivity of NO molecule was over 4.5 times higher than oxygen gas molecule at around 500°C (14, 16). Taking advantages of the high NO selectivity in the self-assembled catalytic layer for a sensing electrode, we evaluated the gas sensing properties of the electrochemical cell with simple arrangement of electrodes.

EXPERIMENTAL

The electrochemical cell with an improved configuration of multilayer sensing electrode (ML-SE) for detection of NO gas in the presence of excess O_2 can be represented by the following cell arrangement:

$$2YSZ \mid Pt\text{-}8YSZ \mid NiO\text{-}8YSZ \text{ (Sensing layer)} \mid Electrolyte \mid Pt\text{-}8YSZ. \qquad [1]$$

8YSZ (8mol% Y_2O_3 doped ZrO_2) disk (20mm in diameter and 0.5mm thick) was selected as a solid electrolyte of the cell. The mixed oxide paste of NiO/8YSZ with 50 mol% of NiO was screen-printed on the surface of 8YSZ electrolyte disk as the sensing layer (SL: 1.77 cm^2), and sintered at 1400 °C for 3 h. The sintered SL should be fully dense structure. Then the Pt-8YSZ composite paste with 70 vol% of Pt was screen-printed to prepare a net-shaped electrode over the SL and calcined at 1150 °C for 1 h. Finally, the 2YSZ (2mol% Y_2O_3 doped ZrO_2) paste was screen-printed as the cover layer over the current collector and sintered at 1450 °C for 1 h. This cover layer increases the reduction rate of NiO to Ni, in consequence, the large amount of Ni grains generate in the NiO/8YSZ interface region by suppressing of the O_2 gas adsorption and decomposition on the open surface of NiO/8YSZ at the top of sensing layer (14). The Pt-8YSZ composite paste with 45 vol% of Pt was screen-printed as a counter electrode (CE) of the cell on to the other surface of electrolyte disk and then sintered at 1400 °C for 1 h.

Electrochemical cell was set in the quartz tube placed in the furnace and connected to a digital multimeter (Keithley 2010) or a pA meter DC voltage source (Keithley 487), both of which were fitted with an IEEE interface to permit automatic data logging via a computer. The sensing performances of both mixed-potential-type and amperometric-type gas sensor was evaluated in the temperature range of 350–500 °C. The sample gases containing various concentrations of NO were prepared by diluting parent standard gases (3000 ppm NO in N$_2$) with dry N$_2$ and O$_2$. The total flow rate of the sample gas or the base gas was adjusted at 300 ml/min. Mass flow controllers were employed to obtain an accurate gas mixture of NO, O$_2$ and N$_2$. The difference in potential (EMF) between ML-SE and CE was measured by means of a digital multimeter as a sensing signal of the mixed-potential-type sensor when the cell was exposed to the base gas or to the sample gas with different NO$_X$ concentrations. For amperometric measurements, the range of applied voltage to the electrochemical cell was established between 0.05 and 0.2V. The concentrations of NO and O$_2$ in the sample gas were monitored using an on–line NO$_X$ (NO, NO$_2$ and N$_2$O) gas analyser (Best Instruments BCL-100uH, BCU-100uH) and a gas chromatograph (CHROMPACK Micro-GC CP 2002), respectively.

RESULT AND DISCUSSION

To clarify the capability of the electrochemical cell with an improved configuration of multilayer sensing electrode as a NO$_X$ sensor, we carried out the measurements of transient response of EMF and electric current at fixed temperature (350–550 °C). The EMF response of the cell at 400 and 450 °C for 5–300 ppm NO in 4%O$_2$ is exemplified in Fig. 1. The EMF values of the cell fabricated were close to zero when NOx was absent in the carrier gas. Therefore, the measured EMF values were regarded to the NO sensitivities. As the temperature increased, the absolute EMF value at the same concentration of NO decreased. The EMF values were almost linear to the logarithm of NO concentration at each temperature examined, and the direction of the EMF response was negative to NO. The EMF magnitude was about -43 and -36 mV upon exposure to 100 ppm of NO at 400 and 450 °C, respectively. The EMF responses in these cases seem to be also based on mixed potential. At the higher temperature range, the comparable results have already been reported by other researchers (17-21).

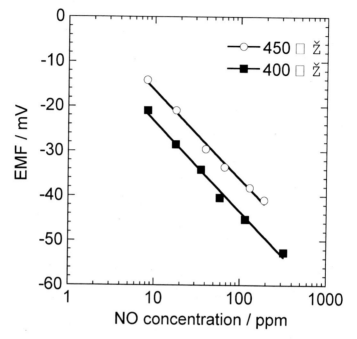

FIGURE 1. Dependence of EMF values on the logarithm of NO concentration for the electrochemical cell with an improved configuration of multilayer sensing electrode at 400 and 450 °C for 5–300 ppm NO in 4%O$_2$.

Figure 2 shows the EMF response and recovery transients to NO in air for the cell at 450 °C. The EMF response of this cell was stable and reproducible. However, in spite of high sensitivity to NOx in the temperature range of 350–550 °C, the response and the recovery times of the cell were rather slow. The lower is the concentration of NO, the longer are the response and the recovery times of the cell. In addition, the cell showed the positive EMF responses immediately after the exposure of NO. When the concentration of NO gasses is lower, these positive EMF responses become smaller and longer. It is expected that the signal originates in the adsorption of NOx and the adsorptions of O$_2$ and/or N$_2$ on the surface of electrodes.

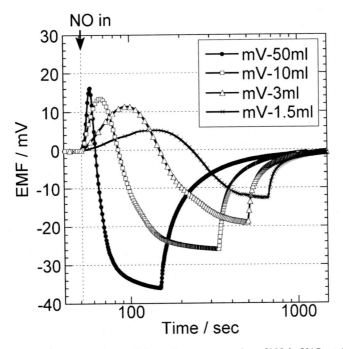

FIGURE 2. Transient response of the cell for various concentration of NO in 2%O$_2$ at 450 °C

When −0.07V (versus CE) was applied to ML-SE to make the cathodic reduction of NO, the current response of the cell at 450 °C for 0–350 ppm NO in 2%O$_2$ is exemplified in Fig. 3. The current value of the cell as fabricated was -1.69x10^{-6}A when NO was absent in the carrier gas. The reproducibility was quiet good when the concentration of NO was changed. The current values were almost linear to the logarithm of NO concentration at 450 °C. But, there is an inflection point at around 20ppm of the NO concentration. This cell shows a high NO sensitivity of over 33nA/ppm at especially less than 20ppm of the NO concentration. In the low NO concentration region, it is supposed that resistance is high because of the short supply of NO.

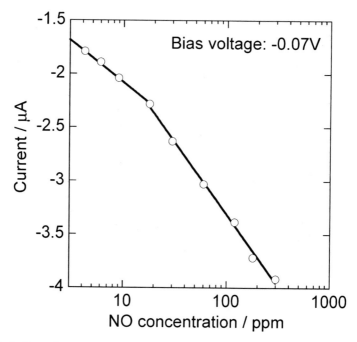

FIGURE 3. Dependence of current values on the logarithm of NO concentration for the electrochemical cell with an improved configuration of multilayer sensing electrode at 450 °C for 5–350 ppm NO in 2%O$_2$.

Figure 4 shows the comparison of transient response of mixed-potential-type and amperometric-type gas sensor for 100ppm NO in 2%O$_2$ at 450 °C. The 90% response was defined to be the time for reaching the 90% steady response current or EMF by switching the gas inflow from 0 to 100 ppm NO. The 10% steady response current or EMF was defined to be the recovery time when the concentration of NO in the gas inlet was switched from 100 to 0 ppm. The EMF response and recovery times were found to be about 1.5 and 6 min, respectively. The current response and recovery times were found to be about 0.25 and 0.7 sec, respectively. The cell showed that the current response of NO sensitivity was about 6 times faster than the EMF response for 100ppm NO in 2%O$_2$ at 450 °C. Fast response is a very important characteristic for a practical NO$_X$ sensor.

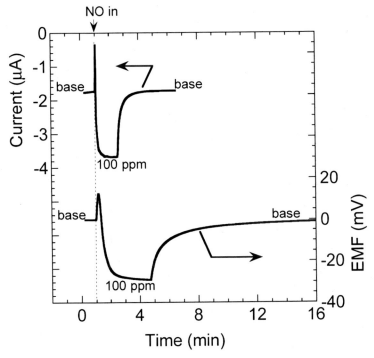

FIGURE 4. Comparison of transient response of mixed-potential-type and amperometric-type gas sensor for 100ppm NO in 2%O$_2$ at 450 °C

CONCLUSIONS

The mixed-potential type and the amperometric-type NOx sensors using the electrochemical cell with an improved configuration of multilayer sensing electrode were fabricated and examined for the sensing properties in the temperature range 400–500 °C. As a result, the cell was found to give the high sensitivity of NO. The magnitude of EMF response was about -43 mV upon exposure to 100 ppm of NO at 400 °C. The amperometric-type gas sensor shows a high NO sensitivity of 33nA/ppm at especially less than 20ppm of the NO concentration. The current response and recovery times were found to be about 0.25 and 0.7 min, respectively. The availability of the self-assembled catalytic layer for NOx sensing and decomposition was confirmed. It behavior opens up possibilities for the development of highly sensitive NO gas sensors, i.e. the NO concentration in the presence of excess O$_2$ can be directly measured without the two-chamber O_2-evacuation system. Moreover, the development of the electrochemical cell system that integrates the function of detection and purification of NO_X is expected.

REFERENCES

[1] A. Fritz, V. Pitchon, *Appl. Catal.*, **B 13** (1997) 1–25.

[2] K. Tamaru, G.A. Mills, *Catal. Today*, **22** (1994) 349–360.

[3] M. Shelef, Catal. *Rev. Sci. Eng.*, **11** (1975) 1–40.

[4] M. Shelef, *Chem. Rev.*, **95** (1995) 209–225.

[5] M. Koebel, M. Elsener, G. Madia, *Ind. Eng. Chem. Res.*, **40** (2001) 52–59.

[6] V.I. P^arvulescu, P. Grange, B. Delmon, *Catal. Today*, **46** (1998) 233–316.

[7] N. Kato, K. Nakagaki, N. Ina, *SAE Paper*, 960334 (1996).

[8] N. Kato, Y. Hamada, H. Kurachi, *SAE Paper*, 970858 (1997).

[9] N. Kato, H. Kurachi, Y. Hamada, *SAE Paper*, 980170 (1998).

[10] H. Inagaki, T. Oshima, S. Miyata, N. Kondo, *SAE Paper*, 980266 (1998).

[11] N. Kato, N. Kokune, B. Lemire, T. Walde, *SAE Paper*, 1999-01-0202 (1999).

[12]S. Bredikhin, K. Maeda, M. Awano, *J. Electrochem. Soc.*, **148** (2001) D133-D138 .

[13]K. Hamamoto, Y. Fujishiro, M. Awano, S. Katayama, S. Bredikin, *Mater. Res. Soc. Symp. Proc.*, **835** (2005) K9.1.1.

[14]K. Hamamoto, T. Hiramatsu, O. Shiono, S. Katayama, Y. Fujishiro, S. Bredikhin, M. Awano, *J. Ceram. Soc. Japan*, **112** (2004) S1071.

[15]K. Hamamoto, Y. Fujishiro, M. Awano, *Solid State Ionics,* **177,** (2006) 2297–2300.

[16]K. Hamamoto, Y. Fujishiro, M. Awano, *J. Electrochem. Soc.*, **153** (2006) D167-D170.

[17] N. Miura, G. Lu, N. Yamazoe, H. Kurosawa, M. Hasei, *J. Electrochem. Soc.* **143** (1996) L33.

[18] N. Miura, G. Lu, N.Yamazoe, *Sens. Actuators*, **B52** (1998) 169.

[19] G. Lu, N. Miura, N. Yamazoe, *Ionics* **4** (1998) 16.

[20] S. Zhuiykov, T. Ono, N. Yamazoe, N. Miura, *Solid State Ionics* **152– 153** (2002) 801.

[21] E. D.Bartolomeo, N. Kaabbuathonga, A. D'Epifanioa, M. L. Grillia, E. Traversaa, H. Aonob, Y. Sadaokab, *J. Euro. Ceram. Soc.*, **24** (2004) 1187.

THE STRUCTURE, ELECTRICAL AND CO-SENSING PROPERTIES OF PEROVSKITE-TYPE $LA_{0.8}PB_{0.2}FE_{0.8}CU_{0.2}O_3$ CERAMIC

Peng Song[a] *, Mingliang Shao[a], Xijian Zhang[b]

[a] Department of Materials Science and Engineering, University of Jinan, Jinan, 250022, P.R.China
[b] Department of Physics and microelectronics, Shandong University, Jinan, 250100, P. R. China

ABSTRACT

Nanocrystalline $La_{0.8}Pb_{0.2}Fe_{0.8}Cu_{0.2}O_3$ has been prepared by the citrate method. The structure, electrical and CO gas-sensing properties were investigated. The results demonstrated that $La_{0.8}Pb_{0.2}Fe_{0.8}Cu_{0.2}O_3$ is a perovskite phase with the orthorhombic structure, it shows p-type semiconducting properties and it has good sensitivity to CO gas.

1. INTRODUCTION

Since environmental protection and domestic ambient atmosphere detection in many countries have recently come to attention, the development of solid state gas sensors, especially for quantification of toxic gases in the air, is more important. Among these gases, CO is more interesting for study, because this highly toxic gas can attach to the haemoglobin, which damages the human body by producing a reduction in cellular respiration [1].

Recently, for CO sensor development, different ceramic materials formulations of the perovskite structure (ABO_3) in the solid state CO sensors have been studied [2-4]. Sensors based on the ABO_3-type composite oxides materials have an advantage of high stability. The sensitivity and selectivity of this kind of sensors can be controlled by selecting suitable A and B atoms or chemical doping as $A_{1-x}A_x \cdot B_{1-x}B_x \cdot O_3$ materials [5-11]. C. M. Chiu and Y. L. Chai have reported that perovskite type $La_{0.8}Sr_{0.2}Co_{1-x}Ni_xO_{3-\delta}$ and $La_{0.8}Sr_{0.2}Co_{1-x}Cu_xO_{3-\delta}$ films showed high sensitivity to CO in a low temperature range.

We recently found that Pb-doping could improve the sensitivity, selectivity and response time of $LaFeO_3$, and $La_{0.8}Pb_{0.2}FeO_3$ calcined at 800 °C showed the best performance [12, 13]. For the reasons mentioned above, we studied this kind of material based on $La_{0.8}Pb_{0.2}Fe_{1-x}Cu_xO_3$ perovskite structure for CO detection. The main effort was directed at the relationship of structure, electric properties and sensing properties to CO with Cu-doping. It's also shown that maximum sensitivity can be obtained for $La_{0.8}Pb_{0.2}Fe_{0.8}Cu_{0.2}O_3$ sintered at 800°C. In this paper, $La_{0.8}Pb_{0.2}Fe_{0.8}Cu_{0.2}O_3$ powder was prepared by the sol-gel method and some electrical and CO sensing properties were investigated.

2. EXPERIMENTAL

---

* Corresponding author
E-mail: mse_songp@ujn.edu.cn

2. 1 Preparation of $La_{0.8}Pb_{0.2}Fe_{0.8}Cu_{0.2}O_3$ powder

The nanocrystalline $La_{0.8}Pb_{0.2}Fe_{0.8}Cu_{0.2}O_3$ powder was prepared by a citric method: firstly $La(NO_3)_3$, $Fe(NO_3)_3$, $Pb(NO_3)_2$, $Cu(NO_3)_2$ and citric acid (all analytically pure) were dissolved in ion-free water at 70°C. Then polyethylene glycol (PEG molecular weight 20, 000) was added under constant stirring to obtain a sol, and the sol was dried into a gel. The gel pieces were ground to form a fine powder. For subsequent annealing, the samples were placed in an oven at 800°C for 3 h. The powder constituents were characterized by X-ray diffraction (XRD) study (RIGAKUD-MAX-γA).

2. 2 Fabrication and measurement of sensors

The prepared $La_{0.8}Pb_{0.2}Fe_{0.8}Cu_{0.2}O_3$ powders were mixed with a polyvinyl acetate (PVA) solution and then ground into paste. Then, the paste was coated onto an Al_2O_3 tube on which two electrodes had been installed at each end. The Al_2O_3 tube was about 8 mm in length, 2 mm in outer diameter, and 1.6 mm in inner diameter. In order to improve their stability and repeatability, the gas sensors were calcined at 400°C for 2 h.

Because $La_{0.8}Pb_{0.2}Fe_{0.8}Cu_{0.2}O_3$ is a p-type material, the electrical resistance of the sensors increases in a reducing gas circumstance. The sensitivity S was defined as: $S = (R_g - R_a)/R_g \times 100(\%)$, where R_g was the DC (direct current) resistance measured under a working circumstance (air with different concentration of CO), while R_a was the resistance in air. The sensor resistance was measured by using a conventional circuit in which a sensor element was connected with an external resistor in series at a circuit voltage of 2.5 V.

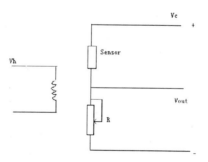

Fig. 1. Graphic of measuring principle.

The scheme of the circuit was shown in Fig. 1. The heating voltage (V_h) was supplied to either of the coils for heating the sensors and the circuit voltage (V_c) was supplied across the sensors and the load resistor (R) connected in series. The signal voltage (V_{out}) across the load, which changed with sort and concentration of gas, was measured. The resistance of the load resistor was alternative and the heating voltage can be adjusted within a wide range.

3. RESULTS AND DISCUSSION

3. 1 Phase composition and structure of ultrafine powders

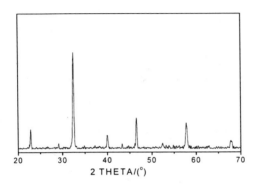

Fig. 2 The XRD patterns of $La_{0.8}Pb_{0.2}Fe_{0.8}Ni_{0.2}O_3$ powders.

Fig. 2 shows the XRD patterns of $La_{0.8}Pb_{0.2}Fe_{0.8}Cu_{0.2}O_3$ powders prepared by the citric method calcined at 800°C for 3h. It is found the sample was perovskite phase with orthorhombic structure and no other phase was observed in the patterns.

The lattice parameters of the samples were calculated from XRD patterns based on:

$$d = (h^2 / a^2 + k^2 / b^2 + l^2 / c)^{-1/2} \tag{1}$$

Where (h,k,l) is the indices of crystallographic plane, d is the interplanar distance, (a,b,c) is the lattice parameters. And the particle sizes (D) were measured from XRD peaks based on the Scherrer's equation:

$$D = 0.89\lambda / (\beta cos\theta) \tag{2}$$

where λ is the wavelength of X-ray, θ the diffraction angle and β the true half-peak width. It is estimated that the value of D is about 31.26 nm and the lattice parameters $a = 5.5224$ Å, $b = 5.5132$ Å, $c = 7.7982$ Å.

3. 2 Electrical conductivity

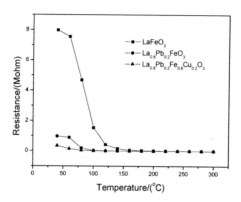

Fig. 3   The resistance-temperature relationships of the samples.

Fig. 3 shows the behaviour of resistance values against temperature in the samples. In the whole temperature range, the resistance values of all sensors reduce with increasing temperature, which is the intrinsic characteristic of semiconductor. The change in resistance with temperature is considerably reduced by Pb-doping and Cu-doping. LaFeO$_3$ is a kind of p-type semiconductive material. Using Kroger-Vink defect notations, its charge carriers are holes (h$^\cdot$), which are produced by the ionization of the La^{3+} cation vacancy defect [V$_{La}^{\times}$]:

$$V_{La}^{\times} \rightarrow V_{La}^{\prime\prime\prime} + 3\,h^\cdot \tag{3}$$

When La^{3+} in LaFeO$_3$ is replaced by Pb^{2+} at A-site and Fe^{3+} is replaced by Cu^{2+} at B-site, the carrier concentration of the compounds will depend on the holes produced by the ionization of [Pb$_{La}^{\times}$] and [Cu$_{Fe}^{\times}$]:

$$Pb_{La}^{\times} \rightarrow Pb_{La}^{\prime} + h^\cdot \tag{4}$$

$$Cu_{Fe}^{\times} \rightarrow Cu_{Fe}^{\prime} + h^\cdot \tag{5}$$

In the formulas, Pb$_{La}^{\times}$ and Cu$_{Fe}^{\times}$ mean the point defect, which is produced when Pb^{2+} occupies the sites of La^{3+} and Cu^{2+} occupies the sites of Fe^{3+} in the crystal. Upon the addition of Pb^{2+} and Cu^{2+}, holes (h$^\cdot$) will be generated based on this equation. As a result of [Pb$_{La}^{\times}$], [Cu$_{Fe}^{\times}$]$>>$[V$_{La}^{\times}$], the concentration of h increases, which results in the increasing of conductivity and reducing of resistance.

3. 3 CO gas sensing properties of $La_{0.8}Pb_{0.2}Fe_{0.8}Co_{0.2}O_3$-based sensor

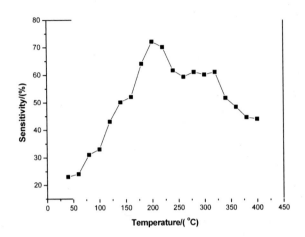

Fig. 4 The sensitivity to 200 ppm CO of $La_{0.8}Pb_{0.2}Fe_{0.8}Cu_{0.2}O_3$-based sensor at different temperature.

Fig. 4 shows the dynamic curve of sensitivity to 200 ppm CO of $La_{0.8}Pb_{0.2}Fe_{0.8}Cu_{0.2}O_3$-based sensor at different operating temperature. According to the dynamic curve it is shown that $La_{0.8}Pb_{0.2}Fe_{0.8}Cu_{0.2}O_3$-based sensor exhibits high sensitivity to CO gas with a wide temperature range of 150-350°C and the optimum operating temperature is about 200°C.

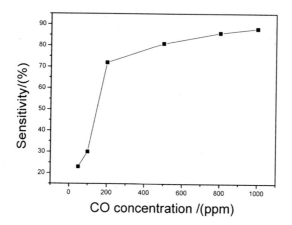

Fig. 5 The relationship between CO concentration and the sensitivity of $La_{0.8}Pb_{0.2}Fe_{0.8}Cu_{0.2}O_3$-based sensor.

When the concentration of CO gas increase, the response of $La_{0.8}Pb_{0.2}Fe_{0.8}Cu_{0.2}O_3$-based sensor also increases. The dependence of the response on the concentration of CO gas at $200°C$ is shown in Fig. 5. The resistance increases significantly with the increase of CO gas concentration, the data can be shown to follow a power law, which is used to describe the sensing characteristics of SnO_2-based materials and can be written as [8,10,14]:

$$R = KC_{CO}^{\alpha} \tag{7}$$

Where R is the resistance of the sensor, C_{CO} is the concentration of CO gas and K and α are constants. Fig. 6 gives the change of electric resistance of $La_{0.8}Pb_{0.2}Fe_{0.8}Cu_{0.2}O_3$-based sensor to CO concentration. In our investigation, this sensor was highly sensitive to CO in a wide temperature range, and the optimum operating temperature was about $200\ °C$, so that the resistance values were obtained at an operating temperature of $200°C$. In comparison, for SnO_2-based material, α is in the range between 1/6 and 1/2 [15,16], while for $La_{0.8}Pb_{0.2}Fe_{0.8}Cu_{0.2}O_3$, the α value is larger than that of SnO_2-based sensors. According to Fig 8, it is estimated that the value of α is about 0.786. It means that the change of electric resistance of $La_{0.8}Pb_{0.2}Fe_{0.8}Cu_{0.2}O_3$-based sensor is strongly affected by the increase of CO gas concentration.

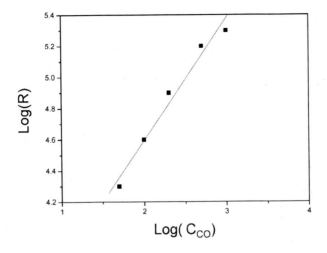

Fig. 6 Log resistance versus Log CO concentration of $La_{0.8}Pb_{0.2}Fe_{0.8}Cu_{0.2}O_3$-based sensor at an operation temperature of $200^{\circ}C$.

4. CONCLUSIONS

The CO gas-sensing material $La_{0.8}Pb_{0.2}Fe_{0.8}Cu_{0.2}O_3$ was synthesized by citric method with calcing temperature $800^{\circ}C$ for 3 h. A perovskite phase with the orthorhombic structure can be obtained. From the investigation of electronic properties, we found that $La_{0.8}Pb_{0.2}Fe_{0.8}Cu_{0.2}O_3$ was a p-type semiconductive material. Furthermore, the change in resistance of $La_{0.8}Pb_{0.2}Fe_{0.8}Cu_{0.2}O_3$-based sensor was considerably reduced by Pb-doping and Cu-doping. And $La_{0.8}Pb_{0.2}Fe_{0.8}Cu_{0.2}O_3$-based sensor show high sensitivity to CO gas with a wide temperature range of 150-$350^{\circ}C$, the experimental results indicate the $La_{0.8}Pb_{0.2}Fe_{0.8}Cu_{0.2}O_3$-based sensor is a good new gas-sensing material for detecting CO gas.

REFERENCES

[1] C. M. Chiu, Y. H. Chang, The structure, electrical and sensing properties for CO of the $La_{0.8}Sr_{0.2}Co_{1-x}Ni_xO_{3-\delta}$ system, Materials Science and Engineering A 266 (1999) 93.
[2] N. N. Toan, S. Saukko, V. Lantto, Gas sensing with semiconducting perovskite oxide $LaFeO_3$, Physica B 327 (2003) 279.
[3] Lorenzo Malavasi, Cristina Tealdi, Giorgio Flor, Gaetano Chiodelli, Valentina Cervetto, Angelo Montenero, Marco Borella, $NdCoO_3$ perovskite as possible candidate for CO-sensors: thin films synthesis and sensing properties, Sensors and Actuators B 105 (2005) 407.

[4] M. S. D. Read, M. S. Islam, G. W. Watson, F. King, F. E. Hancock, Defect chemistry and surface properties of LaCoO$_3$, J. Matter. Chem. 10 (2000) 2298.

[5] H. J. Jung, J. -T. Lim, S. H. Lee, Y. -R. Kim, J. -G. Choi, Kinetics and mechanisms of CO oxidation on Nd$_{1-x}$Sr$_x$CoO$_{3-y}$ catalysts with static and flow methods, J. Phys. Chem. B 100 (1996) 10243.

[6] D. H. Kim, K. H. Kim, Kinetics and oxygen vacancy mechanism of the oxidation of carbon-monoxide on perovskite Nd$_{1-x}$Sr$_x$CoO$_{3-y}$ solutions as a catalys, Bull. Kor. Chem. 15 (1994) 616.

[7] Xiutao Ge, Yafei Liu, Xingqin Liu, Preparation and gas-sensitive properties of LaFe$_{1-y}$Co$_y$O$_3$ semiconducting materials, Sensors and Actuators B 79(2001) 171.

[8] Y. L. Chai, D. T. Ray, H. S. Liu, C. F. Dai, Y. H. Chang, Characteristics of La$_{0.8}$Sr$_{0.2}$Co$_{1-x}$Cu$_x$O$_{3-\delta}$ film and its sensing properties for CO gas, als Science and Engineering A 293 (2000) 39.

[9] C. M. Chiu, Y. H. Chang, The influence of microstructure and deposition methods on CO gas sensing properties of La$_{0.8}$Sr$_{0.2}$Co$_{1-x}$Ni$_x$O$_{3-\delta}$ perovskite films, Sensors and Actuators B 54(1999) 236.

[10] C. M. Chiu, Y. H. Chang, Characteristics and sensing properties of dipped La$_{0.8}$Sr$_{0.2}$Co$_{1-x}$Ni$_x$O$_{3-\delta}$ film for CO gas sensors, Thin Solid Films 342 (1999) 15.

[11] Y. L. Chai, D. T. Ray, G. J. Chen, Y. H. Chang, Synthesis of La$_{0.8}$Sr$_{0.2}$Co$_{0.5}$Ni$_{0.5}$O$_3$ thin films for high sensitivity CO sensing material using the Pechini process, Journal of Alloys and Compounds 333 (2002) 147.

[12] Peng Song, Hongwei Qin, Ling Zhang, Kang An, Zhaojun Lin, Jifan Hu, Minhua Jiang, The structure, electrical and ethanol-sensing properties of La$_{1-x}$Pb$_x$FeO$_3$ perovskite ceramics with x $\leq$ 0.3, Sensors and Actuators B 104 (2005) 312.

[13] Peng Song, Hongwei Qin, Xing Liu, Shanxing Huang, Jifan Hu, Minhua Jiang, Structure, electrical and CO-sensing properties of the La$_{0.8}$Pb$_{0.2}$Fe$_{1-x}$Co$_x$O$_3$ system, Sensors and Actuators B 119 (2006) 415-418.

[14] Y. C. Chen, Y. H. Chang, G. J. Chen, Y. L. Chai, D. T. Ray, The sensing properties of heterojunction SnO$_2$/La$_{0.8}$Sr$_{0.2}$Co$_{0.5}$Ni$_{0.5}$O$_3$ thin-film CO sensor, Sensors and Actuators B 96 (2003) 82.

[15] J. Watson, K. Ihokura, G. S. Coles, The tin dioxide gas sensor, Meas. Sci. Technol. 4 (1993) 711.

[16] P. K. Clifford, D. T. Tuma, Characteristics of semiconductor gas sensors. I. Steady state gas response, Sensors and Actuators B 3 (1982) 233.

$NiCr_2O_4$ AND NiO PLANAR IMPEDANCE-NOx SENSORS FOR HIGH TEMPERATURE APPLICATIONS

M. Stranzenbach
German Aerospace Centre, DLR
Institute for Materials Research
Linder Hoehe
Cologne, Germany, 51147

B. Saruhan
German Aerospace Centre, DLR
Institute for Materials Research
Linder Hoehe
Cologne, Germany, 51147

ABSTRACT

This study uses an innovative type of sensor configuration which contains of an electrolyte constructed of quasi-single crystalline columns, a porous thin layer of sensing electrode (SE) and a conductive Pt reference electrode (RE) deposited on the backside. As electrolyte EB-PVD manufactured discs of PYSZ composition were used. A $NiCr_2O_4$- SE of 3 µm or a NiO- SE of 2 µm thickness were reactive-sputtered on the SE side of the substrate. Gas sensing characterization was carried out in a specially constructed apparatus using different gas concentrations and a mixture of typical flue gases. Impedance and potential analysis were done over a frequency range of 100 kHz - 0.005 Hz and a maximum AC amplitude of 200mV, respectively.

The experiments were carried out between 500 and 750°C under argon as base gas. The sensors with NiO SE were more sensitive towards NO than the sensors with $NiCr_2O_4$-SE at higher temperatures and in net-oxidizing atmospheres. All tested sensor elements were able to detect NO concentrations between 50- 500 ppm and delivered satisfactory results at temperatures up to 550°C. The highest sensing temperature achieved in this study was 600°C. In order to determine the applicability of the sensor, the cross-selectivity towards other flue gases like O_2, CO, CO_2 and CH_4 were also analysed. Both sensors showed a slight change in sensing signal for both oxidizing and reducing gases. The NiO- sensor was successfully tested to detect NO in a wider concentration range under the presence of 5%vol. O_2 at its maximum operating temperature.

INTRODUCTION

Recently, an urgent need for high performance gas sensors has arisen. Especially in the automotive sector, where engines are more and more operated in lean burn areas, the flue gas compositions have changed to net-oxidizing conditions. Furthermore, the monitoring of catalytic converters in the OBD-system gained a central position in the last years. Therefore, the development of new catalysts for the reduction of NO_x is necessary, which makes a permanent sensor control ineluctable. The requirements for these new generation sensors are high temperature stability and long lifetimes. The temperatures for closed coupled catalysts can rise temporarily up to 900°C to which the sensor needs to withstand. A major expectation from such sensors is to be able to monitor total NO_x in the exhaust gas under the presence of oxygen at working temperatures between 500 and 700°C. Besides this total NO$_x$ (NO and NO$_2$) needs to be

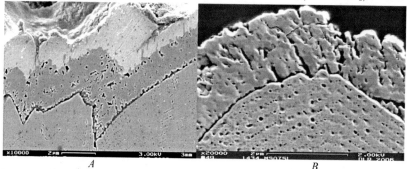

A   B

Figure 1. SEM picture of the NiCr$_2$O$_4$- (A) and the NiO- (B) SE/ Electrolyte- Interface showing a SE average grain size of 0.2 μm (A) and 0.4 μm (B). In the case of (A) the top layer is the porous Pt- collector.

monitored in a concentration range of 0- 500 ppm at adequate response times of less than 10 seconds, for instance, in order to dosage an reducing agent. Standard potentiometric sensors suffer under the opposite sign for the emf potential of NO and NO$_2$ which makes it very hard to monitor total NO$_x$ [1]. To overcome these problems different design changes and sensor types have been suggested (e.g. multi chamber amperometric sensors) [2,3]. But the need for cost-efficient and simple sensor systems with high lifetime holds back their commercial use until today. Therefore, we suggest a novel planar NO$_x$-sensor type, which needs no reference gas electrode, having a magnetron sputtered metal-oxide sensing electrode (SE), for use at high temperature and in harsh environments such as turbo engine vehicles and aircraft turbines [4,5].

EXPERIMENTAL

In this study, an innovative type of sensor configuration is used which consists of an electrolyte constructed of quasi-single crystalline columns, a porous thin layer of sensing electrode (SE) and in the case of NiCr$_2$O$_4$ a thin porous Pt- collector layer on top and a conductive Pt reference electrode (RE) deposited on the backside. As electrolyte EB-PVD manufactured discs of PYSZ composition were used. The discs were 12.7 mm in diameter and about 500 μm thick. The special columnar microstructure of the electrolyte features a smooth RE side and a high surface area SE side. Its preferentially oriented crystal direction lies vertically to the electrodes. For the EB-PVD fabrication of the electrolytes it was necessary to coat the PYSZ on a stable metallic, high temperature substrate, which was afterwards removed by a wet-etching process. After a cleaning step, a short aging procedure of the electrolytes has been performed which was necessary to extend the sensor lifetime and optimise its performance. After this procedure, either a NiCr$_2$O$_4$-SE of approx. 3 μm thickness or a NiO-SE of 2 μm thickness was reactively sputtered on the SE side of the substrate, using O$_2$ as reactive gas (Figure 1). Annealing up to 1000°C in air led to the formation of a porous SE with an average grain size of approximately 0.2 μm and 0.4 μm respectively (Figure 1). Following this, thin and porous Pt layers were sputtered on the RE and in the case of the NiCr$_2$O$_4$ as well on the SE side serving as

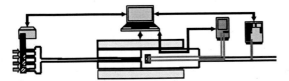

Figure 2: Schematic drawing of the experimental setup. The sensor is connected to the impedance spectroscope at the rear part of the furnace (right side of the figure)

ultra-thin electrodes or electron collector. Finally the sensor was electrically connected using a commercial Pt-paste (LPA 88/11S, Hereaus) and 0.15 mm Pt wires which were hardened at 1000°C. Gas sensing characterizations were carried out in a specially constructed apparatus consisting of a tube furnace and a custom-built quartz glass reactor (Figure 2).

The gas mixtures were controlled by MFCs and adjusted at a total flow of 200 sccm/min. NO$_x$ concentrations were varied between 0 and 1000 ppm. Cross-selectivity was analyzed towards O$_2$ (0–20 vol.%), CH$_4$ (0-500 ppm), CO (0-400 ppm) and CO$_2$ (0-500 ppm). Argon was used as balance gas in all cases. Impedance and potential analysis were carried out by a Solartron 1255b and a Solartron 1286. Impedance spectroscopy was done at a frequency range of 100 kHz to 0.005 Hz and maximum AC amplitude of 200 mV. No DC Bias was applied. Morphological analysis in terms of porosity, grain size and surface condition was done by means of SEM (LEO 982). XRD-analysis for both cases was carried out by means of a Siemens D5000 x-ray diffractometer (Cu Kα radiation) to determine the phase sequence and crystal structure.

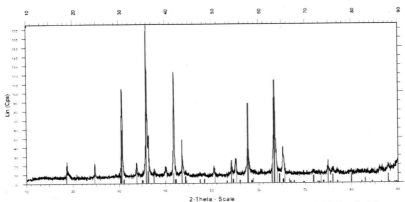

Figure 3. XRD analysis of the sensing electrode after annealing at 1000°C for 4h. The cubic NiCr$_2$O$_4$ phase is the dominating crystal phase. ☐ NiCr$_2$O$_4$ cubic ☐ NiCr$_2$O$_4$ tetragonal ☐ NiCrO$_3$ rhombohedral

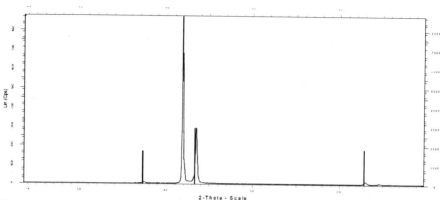

Figure 4. XRD spectrum of the NiO sensing electrode. Marked are the peaks of the pure NiO cubic phase. The none-marked peaks are from the utilised substrate.

Figure 3 shows that the main crystalline phase in the annealed NiCr-oxide sensing electrode layer is a cubic NiCr$_2$O$_4$ spinel phase, aside this phase there are two other crystalline phases (one spinel type and one perovskite type). All other peaks in the spectrum are from the substrate spectrum which is not indicated in the figure. Figure 4 shows the pure cubic NiO phase after the annealing process at 1000°C for 4 hours. All other peaks in the spectrum are also due to the substrate and are not indicated in Figure 4.

RESULTS AND DISCUSSION

The experiments were carried out between 500 and 750°C. Since the thermodynamic equilibrium of NO$_x$ between NO and NO$_2$ is of about 90% on the NO side at 600°C, we used a pure NO gas-mixture for the sensor characterisations. The impedance spectroscopy was done with 126 measuring points over the frequency interval with two measuring cycles for each point. For the first NO sensing behaviour characterisation for each SE material a virgin sensor was used. The Nyquist-plots of the complex impedance showed in both cases (NiO- and NiCr$_2$O$_4$-SE) distinct variations by NO-concentration in the lower frequency range and a decreasing sensor signal to NO with increasing temperature. For the use in net-oxidizing atmospheres the most important cross-selectivity is towards oxygen. Therefore, the sensing behaviour was characterized towards several NO concentrations in oxygen gas mixtures. Figure 5 shows the sensing behaviour towards NO for NiO and NiCr$_2$O$_4$ in the presence and absence of oxygen and at a gas temperature of 600°C. Figure 5 displays the impedance spectra for 100 and 500 ppm NO in the presence and absence of 5 vol.% O$_2$. As it can be seen from Figure 5, NiCr$_2$O$_4$ has a similar selectivity towards NO compared to NiO in the absence of oxygen. But in the presence of 5 vol.% O$_2$ NiO shows much lower cross-selectivity towards O$_2$ and its sensitivity towards NO is maintained. In contrast to this the selectivity of NiCr$_2$O$_4$ towards NO fades away and the signal characteristic (i.e. resistance) is strongly affected. The NiO impedance spectra are also shifted by the presence of O$_2$, but nearly negligible in a very small scale. The sensing characteristics towards NO are, in turn, not affected.

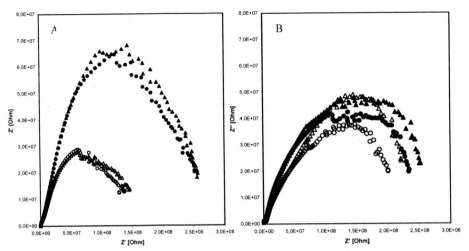

Figure 5. Impedance spectra of NiCr₂O₄ (A) and NiO (B) with 100 ppm (triangular) and 500 ppm (circle) NO in the absence (full symbols) and the presence of 5 vol.% O₂ (open symbols).

NiO and NiCr₂O₄ have a similar impedance spectra, indicating a likewise signal formation. Due to the improved selectivity of NiO, the cross-selectivity towards other typical flue gases were characterized for both NiCr₂O₄ and NiO, expecting an improved sensing behaviour in the gas mixtures for NiO. In comparison to NiCr₂O₄ [6], NiO has a much lower sensitivity towards O₂ and shows no ordered signals to oxygen concentration between 0 and 20 vol%. (Figure 6).

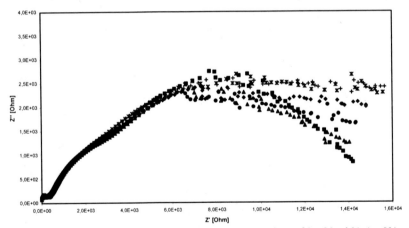

Figure 6. Impedance spectra of NiO with oxygen concentrations of 0 – 20vol.%. (■=0% ▲=2% ●=5% ◆=10% *=15% +=20%)

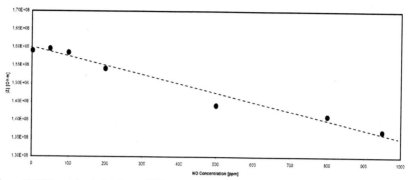

Figure 7. NiO sensing behaviour of |Z| to different NO concentrations in 5vol.% @ 0.1 Hz.

Also typical flue gases such as carbonmonoxide (CO) and UHCs are reducing gases. NiO shows only a baseline shift, in the presence of CH$_4$ as a representative for UHCs and displays hardly any reaction to the presence of CO. NiCr$_2$O$_4$ instead shows a cross-selectivity towards CH$_4$ and CO [6]. The cross-selectivity towards CO and CH$_4$ were also characterised in a gas mixture containing Ar plus 5 vol.% O$_2$. Aas expected , both sensors showed no more sensitivity towards CO and CH$_4$, which is most likely due to the oxidation of CO and CH$_4$ to CO$_2$ at the experiment temperature of 600°C.

For practical applications it is important to have a low cost and fast responsing sensor. Therefore we suggest a sensor with a fixed measuring frequency and to monitor the total impedance |Z| ($|Z| = \sqrt{(Z')^2 + (Z'')^2}$). This step decreases the measuring time from several minutes to a few seconds without information loss. This makes it necessary to choose a measuring point in the low frequency spectrum, where the relevant electrochemical processes for the signal formation can follow the frequency. But this is in conflict with measuring time and therefore with response time. As a good compromise between selectivity and response time, 0.1 Hz was chosen as measuring frequency. It can be seen in Figure 7 that NiO has a linear sensing behaviour towards NO in net-oxidizing atmospheres indicating a non-Nernst sensing behaviour. In Figure 7, the response of the sensor at 0.1 Hz measuring frequency to NO concentrations between 50 and 950 ppm NO is shown. The dashed line displays the linearity of the signal over the whole concentration range at 600°C. Reaction- and regeneration time measurements were also performed over a wide NO-concentration range in 5 vol.% oxygen and at 0.1 Hz frequency. The T$_{90}$ values are in the range of 60 seconds and regeneration times are in the range of 90 seconds. Both times seem to be independent from concentration. As it can be seen from Figure 9, the sensor signal is stabile over a fairly long time and no baseline shift could be observed over several hours.

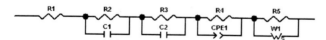

Figure 8. Equivalent circuit for the simulation of the impedance spectra

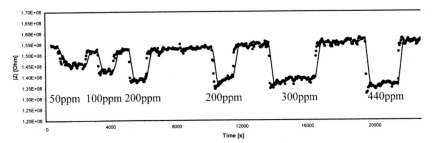

Figure 9. Reaction to 50, 100, 200, 200, 300 and 440 ppm NO with a base gas composition of 5vol.% O$_2$ and argon.

In order to better understand the signal formation and sensing behaviour the impedance spectra were simulated by an equivalent circuit, shown in Figure 8. The results show, that basically for both SE-materials the Cole element varies with NO (and O^{2-}) concentration [7]. As in [8] described, the signal formation is probably due to a double layer charge, most likely caused by a dissociation and the formation of O^{2-} -ions. In the case of NiO the dissociation seems to be driven mainly by NO and NO$_2$ whereas NiCr$_2$O$_4$ seems to be more unselective and also to promote the dissociation of oxygen.

CONCLUSIONS

This study shows the potential of Ni-based oxide-SE impedance-metric sensors as high temperature NO$_x$ sensors for the use in hot exhaust gas applications under harsh environments and the applicability of the total impedance as a reliable sensor signal. It is shown that NiO is a more promising candidate for the application in lean atmospheres and offers lower cross selectivity towards other flue gases and oxygen compared to NiCr$_2$O$_4$. Furthermore NiO has a potential to operate at higher temperatures, when employed with an optimised reference electrode. As shown, both sensors work over a wide range of NO-concentrations (0 – 1000 ppm) at high temperatures up to 600°C. The NiO- SE is able to monitor NO under the presence of O$_2$, but oxygen concentration should be kept constant at the sensing electrode over a certain range. In both cases, cross-sensitivities towards reducing gases were observed in different sensitivities, however, they are not significant during the presence of oxygen. Regarding the aimed service conditions of the sensor, this behaviour should not influence the sensing characteristics under net-oxidizing atmospheres. In the case of NiO, the presence of oxygen seems not to restrict the NO-detection ability. A short analysis of the received impedance spectra and the creation of a capable equivalent circuit led to a satisfactory simulation of the sensor behaviour. It is shown that a Cole element is the main element responsible for the changes in the equivalent circuit due to NO-concentration variations. In combination with the experimental data, the signal formation is in both cases mainly due to the dissociation of NO and O$_2$ respectively. Regarding these results, it can be postulated that Ni-based oxide-SEs are good candidates for SEs in planar flue-gas sensors at high temperatures.

REFERENCES

[1] F. Ménil, V. Coillard, C. Lucat; "Critical review of nitrogen monoxide sensors for exhaust gases of lean burn engines", *Sens. Actuators, B, Chem*, **67**, 1-23 (2000)

[2] J. Schalwig, S. Ahlers, P. Kreisl, C. Bosch-v. Braunmühl, G. Müller; "A solid-state gas sensor array for monitoring NO$_x$ storage catalytic converters", *Sens.. Actuators, B, Chem*, **101**, 63-71 (2004).

[3] W. Göpel, G. Reinhardt, M. Rösch; "Trends in the development of state amperometric and potentiometric high temperature sensors", *Solid State Ionics*, **136-137**, 519-531 (2000).

[4] S. Zhuiykov, T. Nakano, A. Kunimoto, N. Yamazoe, N. Miura; "Potentiometric NO$_x$ sensors based on stabilized zirconia and NiCr$_2$O$_4$ sensing electrode operating at high temperatures", *Electrochemistry Communications*, **3**, 97-101 (2001).

[5] N. Wu, Z. Chen, J. Xu, M. Chyu, S. Mao; Impedance-metric Pt/YSZ/Au-Ga$_2$O$_3$ sensor for CO detection at high temperatures, *Sens. Actuators, B, Chem*, **110**, 49-53 (2005).

[6] E. Gramckow, *Master Thesis*, University of applied Science Bonn-Rhein-Sieg, Bonn (2006)

[7] M. Stranzenbach, E. Gramckow, B. Saruhan, "Potentiometric NO$_x$ sensor based on stabilized zirconia and NiCr$_2$O$_4$ sensing electrode operating at high temperatures", *Sens. and Act. B*, submitted (2006).

[8] M. Stranzenbach, B. Saruhan, "Planar, impedance- metric NO$_x$ sensor with spinel- type SE for high temperature applications", *Proc. Eurosensors '06*, pp. 352-353 (2006).

Thermoelectric Materials for Power Conversion Applications

THE DEVELOPMENT OF THERMOELECTRIC OXIDES WITH PEROVSKITE-TYPE STRUCTURES FOR ALTERNATIVE ENERGY TECHNOLOGIES

A. Weidenkaff, M. Aguirre, L. Bocher, M. Trottmann, and R. Robert

Empa - Swiss Federal Laboratories for Materials Testing and Research
Ueberlandstr.129
CH-8600 Duebendorf
e-mail: anke.weidenkaff@empa.ch

ABSTRACT

Direct and efficient thermoelectric conversion of solar or geothermal waste heat into electricity requires the development of p- and n-type semiconductors with similar materials properties. Perovskite-type transition metal- oxides are investigated as potential candidates for thermoelectric devices operating at high temperatures as they can possess large positive as well as large negative thermopower depending on their composition. The three parameters defining the thermoelectric figure of merit ZT are in most cases interdependent: The thermopower increases with increasing resistivity. The heat conductivity increases with electric conductivity. Therefore an optimum charge carrier concentration and mobility has to be defined, which is depending on e.g. the substitution level, the spin states of the transition metals, the ligand field, i.e. the crystallographic structure, the valence states of the cations, ionic deficiencies, etc. It was further shown that the heat conductivity can be lowered by enhanced boundary scattering in nanostructured oxides without changing the electronic transport properties.

INTRODUCTION

Heat can be directly converted into electricity with thermoelectric devices. The use of geothermal or solar heat as energy source for a thermoelectric generator is an attractive and environmentally clean (CO_2-free) way to generate electrical power [1, 2]. Thermoelectric converters do not depend on mechanical or chemical conversion processes. Thus, they are emission free during operation, noiseless and extremely durable. The amount of electrical power produced is depending on the thermoelectric conversion efficiency of the device and the heat flux.

The direct conversion of the heat flux into electricity is connected to electron transport phenomena, and the interrelated Seebeck effect. Thermoelectric power is defined as the entropy carried by an electron [3].

The direct efficient thermoelectric conversion of solar or geothermal heat into electricity requires the development of p- and n-type semiconductors with similar materials properties. In general compounds exhibiting a considerably large Seebeck coefficient S, high electrical conductivity σ, and a small thermal conductivity κ, in summary a large thermoelectric figure of merit ZT are required (eq.1).

$$ZT = S^2 \sigma T/\kappa \qquad \text{eq.1}$$

Since these transport properties are interconnected by the Wiedemann-Franz-Law for most materials, the development of a material breaking this relationship is a scientific challenge. Classic thermoelectric materials suffer from high toxicity, low stability and low efficiency [4], while ceramics have been recently recognised as good thermoelectrics with high stability even at elevated temperatures and low production costs.

Lately large thermopower was discovered in perovskite-type materials making them promising candidates for future thermoelectric applications [5, 6]. The thermoelectric effect relies on the fact that the flow of electrons in a solid produces entropy current as well as a charge current. In complex transition metal- oxides the spin of electrons can become an additional source of entropy resulting in the large thermopower in these systems [7].

Transition metal- oxides with perovskite structure are potential candidates for thermoelectric devices operating at high temperatures as they can possess large positive as well as large negative thermopower depending on their composition [8-10].

The three parameters defining the thermoelectric figure of merit ZT are in most cases interdependent, i.e. as the thermopower increases so does the resistivity; as the electric conductivity increases so does the heat conductivity. Therefore an optimum charge carrier concentration and mobility has to be defined, which is depending on factors like the substitution level, the spin states of the transition metals, the ligand field, i.e. the crystallographic structure, the valence states of the cations, ionic deficiencies, etc. The heat conductivity can be lowered with suitable substitutions, complex crystal structures or by enhanced boundary scattering in nanostructured materials without changing the electronic transport [11].

Improved thermoelectric materials with p-type conductivity as well as compounds with n-type conductivity have to be developed to produce Thermoelectric Oxide Modules (TOM) - cascades of thermoelectric oxide thermocouples- to convert the applied temperature gradient of geothermal and/or solar heat exchanger systems into electric power. In these devices several ceramic p- and n-legs will be connected electrically in series and thermally in parallel to produce an open circuit voltage of several volts at even low temperature gradients. A reduction of the size of the thermoelectric elements should further serve to increase the specific power (W/cm^2) of the device due to the increased surface area [12].

The LaCoO$_3$ system is a promising thermoelectric material due to its high Seebeck coefficient of 600 µV/K at room temperature. The thermopower of LaCoO$_3$ is positive due to the partial disproportionation $2 Co^{3+} \leftrightarrow Co^{2+} + Co^{4+}$. Nevertheless the electrical resistivity is rather high (10 Ω cm) which lowers the conversion efficiency. The amount of charge carriers and thus the electrical conductivity and thermoelectric properties in this system can be tuned by suitable Co-site and La-site substitution. For the n-type legs manganates phases are considered and prepared as powders.

Polycrystalline powder samples were produced and characterised to study grain size influences on the transport properties.

EXPERIMENTAL

Powders of the Ca- and Ni-substituted substituted La- cobaltates and manganates were synthesised with diverse precursor reactions. The precursors were obtained by dissolving the required amount of La(NO$_3$)$_3$ * 6 H$_2$O (Merck, $\geq$ 97%), Mn(NO$_3$)$_2$ * 6 H$_2$O (Merck, $\geq$ 99%), Ca(NO$_3$)$_2$ * 6 H$_2$O (Merck, $\geq$ 97%) Ni(NO$_3$)$_2$ * 6 H$_2$O (Merck, $\geq$ 97%) and Co(NO$_3$)$_2$ * 6 H$_2$O (Merck, $\geq$ 97%) in water and mixing them with a chelating agent, e.g. citric acid. These complexes were polymerised in a second step and dried to obtain homogeneous aerogel precursors. Phase purity of the products was confirmed by X-Ray Diffraction (XRD) with a PANanalytical X'pert diffractometer using Cu-Kα radiation. The morphology of the calcined powders was studied using a Scanning Electron Microscope (SEM) LEO JSM-6300F with EDX detector. The structure of the samples was further studied by Transmission Electron Microscopy (TEM) using a Philips CM 30. The oxygen content of the powders was determined by the hot gas extraction method using a LECO TC 500.

The transport properties measurements were performed on bar shaped pressed-sintered pellets with general dimensions of 1.65 mm * 5 mm *1 mm. The electrical conductivity and Seebeck coefficient were measured in air simultaneously as a function of temperature from 340 K to 1273 K using a RZ2001i measurement system from Ozawa Science, Japan. The electrical conductivity was determined using a four-point probe method. Two electrical contacts were positioned at both ends of the sample and the two others contacts were on the sample body. Circular pellets (10mm diameter) of the same composition were used for thermal conductivity measurements by the laser flash method using a Netzsch LFA-457 apparatus from 300 K to 1020 K in argon atmosphere.

p- and n-type legs are produced with the obtained samples and assembled to all perovskite ceramic thermoelectric converters.

RESULTS

A series of A and B-site substituted lanthanum cobaltate and manganate compounds with various compositions and perovskite-type structure have been successfully synthesised by a polymeric precursor method [13]. The fine black powders obtained at T < 873 K reveal a uniform particle size of less than 50 nm in diameter. After further annealing to 1273 K, the sintering processes lead to an increase of particle size by a factor of 8. All samples are single phase and crystallize in rhombohedral crystal structure.

The EDX analysis shows a homogeneous composition in agreement with the nominal composition. The oxygen content measurements indicate a slight excess of oxygen for the manganates (δ < 0.06). The cobaltates contain a slight oxygen deficiency (δ < 0.08).

Cobalt ions can appear in low spin-, intermediate spin- and high spin states (see   Table 1), leading to an additional entropy factor - the *spin entropy effect* [7]. This property is most probably the reason for the large thermopower found in complex cobalt oxides. The effects of the strongly interacting or "correlated" electrons in these systems are substantial and lead to exciting related properties [14].

Lanthanum cobaltates show localised to itinerant behaviour of the d-electrons depending on the spin configuration of the cobalt ions [15]. Low spin ions are generally associated with itinerant d- electrons, while high spin ions are associated with localised electrons [16].

Co^{3+}	Co^{4+}	
$t_{2g}^{6}e_{g}^{0}$	$t_{2g}^{5}e_{g}^{0}$	Low Spin
$t_{2g}^{5}e_{g}^{1}$	$t_{2g}^{4}e_{g}^{1}$	Intermediate Spin
$t_{2g}^{4}e_{g}^{2}$	$t_{2g}^{3}e_{g}^{2}$	High Spin

Table 1: Spin states of Co^{3+} and Co^{4+} cations

Factors affecting the thermopower of complex cobalt oxides are:

- the valence of the cobalt atoms,
- the symmetry of the structure,
- the strengths of the crystal field.

Among a wide range of potential perovskite-type candidates for thermoelectric applications produced in our laboratory, we selected metallic 2-D and semiconducting 3-D cobalt oxides to study possibilities for improving the figure of merit $ZT = S^2T/\rho\kappa$ by enhancing the electrical conductivity and the Seebeck coefficient with appropriate cationic substitutions and by lowering the heat conductivity with decreasing the particle size.

With appropriate substitutions the conduction mechanism of Lanthanum cobaltates can be changed from p-type semiconductors to n-type semiconductors, i.e. electron conduction mechanism to hole conduction mechanism respectively. Substitution of trivalent La cations on the A-site position with tetravalent cations (e.g. Ce^{4+}) or substitution in the B-site Co^{3+} matrix with tetravalent cations (e.g. Ti^{4+}) produces mixed valence states of Co^{3+} and Co^{2+}. The combination gives rise to an electron hopping mechanism. If divalent cations are introduced for La at the A-site (e.g. alkaline earth cations) or Co at the B-site (e.g. Ni^{2+}) Co^{4+} can be formed for valence compensation. This leads to a hole hopping transport mechanism. In this way, employing co-doping with di- and tetravalent ions in both A- and B-site positions suitable materials for the p-type and n-type leg of a TOM (see figure 1) can be formed, leading to thermocouples with very similar materials properties.

Figure 1: Thermoelectric Oxide Module (TOM) made of cascades of ceramic thermoelectric p- and n-type elements

The average Co valence can be adjusted with the rate of substitution and by changes in the O sublattice and can be determined from the compositional studies and XRD data on the crystal structure. The substitutions leading to transition metal ions in different oxidation states are important to enhance the electrical conductivity. Generally the effect of increasing spin and orbital degrees of freedom in systems with transition metals in different oxidation states dominates the intrinsic transport and has a positive influence on the thermopower in lightly doped systems. However, with increasing substitution/carrier concentration the Coulomb interactions become less pronounced as in lightly doped systems and the effect of spin and orbital is less important. Although the thermopower is enhanced by orbital degeneracy the mechanism responsible for large thermopower are not always clear [17].

Since changes in the crystallographic structure influence as well the transport properties the interpretation of substitutional effects is rather complex in these systems. The XRD patterns of e.g. the discussed example compounds $LaCo_{1-x}Ti_xO_{3-\delta}$ (x= 0.01, 0.20, 0.30, 0.50) show that the phases change from a rhombohedral to an orthorhombic system with increasing Ti content. Due to the presence of Co^{2+} ions the thermopower of e.g. $LaCo_{0.99}Ti_{0.01}O_{3-\delta}$ is negative with a value of S = -200µV/K at room temperature.

Ni substitution causes reduction of thermopower by a factor of 2 - 3. It was further observed that the Seebeck coefficient is decreasing with rising temperature. This observation is probably due to the formation of Co^{3+} high spin state configuration ($t_{2g}^4 e_g^2$) at high temperatures. In $LaCoO_3$ the Co^{3+} ion can be present in three different spin states. At low temperatures the Co^{3+} ions are in a low spin state ($t_{2g}^6 e_g^0$). With increasing temperature, they undergo transition to an intermediate spin state ($t_{2g}^5 e_g^1$) and at 1200 K to a high spin state configuration ($t_{2g}^4 eg^2$) [18] (see table 1). The thermopower in cobalt oxides can be interpreted with Heikes formula [19]:

$$S = -\frac{\kappa_B}{e} \ln\left(\frac{g_3}{g_4} \frac{x}{1-x}\right) \qquad (eq.1)$$

where x is the carrier concentration and g_3 and g_4 are the degeneracy of configurations of trivalent and tetravalent cobalt in the octahedral coordination, respectively. Applying this formula, large thermopower is expected for low level substituted compounds.

Thus, the absolute value of the thermopower depends on the ratio between Co^{3+} and Co^{4+} configurations and the carrier concentration. Accordingly, as Ni substitution increases the carrier concentration, a reduction of the absolute thermopower value at room temperature results in agreement with Heikes formula [20].

The decrease of the electrical resistivity with increasing Ni content (see figure 2) can be associated to an increase of Co^{4+} and thus the number of holes in the mixed valance band. Hence, the electronic conductivity in the $LaCo_{1-x}Ni_xO_{3-\delta}$ phases increases with increasing Ni content as expected [21].

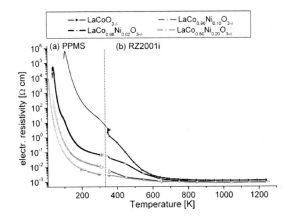

Figure 2: Temperature dependence of the electrical resistivity of $LaCo_{1-x}Ni_xO_{3-}$ (where x = 0.02, 0.08, 0.10, 0.12, 0.20 and 0.30)

The power factor (PF) of the Ni substituted compounds decreases with increasing temperature (see figure 3). However, the compound with x = 0.10 nickel content has a PF = $1.42*10^{-4}$ W/m^2*K at T = 540 K.

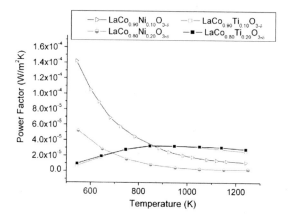

Figure 3: Temperature dependence of the Power Factor of LaCo$_{1-x}$(Ti,Ni)$_x$O$_{3\pm}$ (with x = 0.10, and 0.20) in the temperature range of 500 K < T < 1240K.

The total thermal conductivity of the sub microcrystalline Ni-substituted lanthanum cobaltate compounds is measured in the temperature range of 300 K < T < 1273 K (see figure 4). At room temperature, the values of the thermal conductivity for the produced series of Ni-substituted cobaltates are very small (0.34 - 0.48 W/m K) when compared with the reported values between 1.2 - 2.5 W/m K for cobaltate compounds [22].

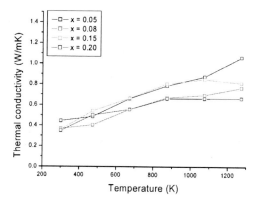

Figure 4: Temperature dependence of the thermal conductivity of LaCo$_{1-x}$Ni$_x$O$_3$ (x = 0.05, 0.08, 0.15, and 0.20) in the range of 300 K to 1273 K.

This observation could be due to the small grain size of the powders leading to a reduction of the lattice conductivity [23]. For all compositions the total thermal conductivity increases slightly with increasing temperature up to 1 W/m K. The lattice thermal conductivity is calculated using the Wiedemann-Franz-Lorenz relation [3]. The obtained value corresponds to 90% of the total thermal conductivity in these systems and is decreasing with increasing Ni content. This leads to the assumption that the heat is carried predominantly by phonons. An increased Ni content in the $LaCoO_3$ system probably causes additional phonon scattering processes.

The specific heat capacity of samples with composition $LaCo_{1-x}Ni_xO_3$ (x = 0.05, 0.08, 0.15, 0.20) is found to be independent of the temperature in the investigated region (300 K < T < 1300 K) ranging between 0.5 and 0.65 J/gK and decreases with increasing Ni content.

The thermopower decreases with increasing Ni content and with increasing temperature. The thermoelectric power factor (PF = $S^2 * \sigma$) at room temperature was optimized for a sample with the composition $LaCo_{0.90}Ni_{0.10}O_{2.9}$. The thermal conductivity for all samples is low (0.34 W/mK < κ < 1 W/mK up to 1200 K) due to the sub micrometer grain morphology of the powders. Consequently, the dimensionless Figure of Merit is improved to a value of ZT = 0.2 at room temperature and remains constant with increasing temperature up to 600 K for samples with Ni contents in the range of 0.08 < x < 0.12. For the $LaCo_{0.80}Ni_{0.20}O_3$ compound an improvement of ZT by a factor of 5 compared with the reported value of ZT = 0.022 has been achieved [24].

The thermal property measurements reveal similar values for the perovskite-type cobaltates and manganates in the whole temperature range from 300 K to 1020 K. The measured total thermal conductivity values the cobaltates vary from 0.5 – 1.8 W/m K while the manganates samples show thermal conductivities between 1.0 W/m K and 1.5 W/m K depending strongly on the particle size (see also [25]).

CONCLUSIONS

The electrical conductivity, the Seebeck coefficient and the thermal conductivity were measured in a broad temperature range. The studied perovskite-type oxides are showing a large potential for thermoelectric applications, as the thermopower relies on the tuneable itinerant electrons. The sign and the absolute value of the Seebeck coefficient can be changed with hole or electron doping i.e. aliovalent cation substitutions.

N-type doping is generated by the Co^{2+}/Co^{3+} mixed valency due to the substitution of Co by Mn or Ti in the $LaCoO_3$ system. With 5 % of Mn in the $LaCoO_3$ system, a negative Seebeck value is obtained at 340 K which indicates predominant electron-like carriers (S_{340K} = -82.1μVK^{-1}).

ACKNOWLEDGEMENTS

The authors thank the Swiss Federal Office of Energy for financial support.

REFERENCES

[1]   D. M. Rowe, "Thermoelectrics, an environmentally-friendly source of electrical power", *Renewable Energy* **16**, 1251 (1999).

[2]   H. Scherrer, L. Vikhor, B. Lenoir, A. Dauscher, P. Poinas, "Solar thermolectric generator based on skutterudites", *Journal of Power Sources* **115**, 141 (2003).

[3]   C. Kittel, *Solid State Physics*, 8 ed., Wiley-VCH, (2004).

[4]   C. M. Bandhari, D. M. Rowe, *CRC Handbook of Thermoelectrics*, CRC Press, Boca Raton, (1995).

[5]   I. Terasaki, Y. Sasago, K. Uchinokura, "Large thermoelectric power in NaCo2O4 single crystals", *Phys.Rev. B* **56**, R12685 (1997).

[6]   A. Maignan, L. B. Wang, S. Hebert, D. Pelloquin, B. Raveau, "Large Thermopower in Metallic Misfit Cobaltites", *Chemistry of Materials* **14**, 1231 (2001).

[7]   Y. Wang, N. S. Rogado, R. J. Cava, N. P. Ong, "Spin entropy as the likely source of enhanced thermopower in NaxCo2O4", *Nature* **423**, 425 (2003).

[8]   A. Maignan, S. Hebert, L. Pi, D. Pelloquin, C. Martin, C. Michel, M. Hervieu, B. Raveau, "Perovskite manganites and layered cobaltites: potential materials for thermoelectric applications", *Crystal Engineering* **5**, 365 (2002).

[9]   R. Fresard, S. Hebert, A. Maignan, L. Pi, J. Hejtmanek, "Modeling of the thermopower of electron-doped manganites", *Physics Letters A* **303**, 223 (2002).

[10]   E. Pollert, J. Hejtmanek, Z. Jirak, K. Knizek, M. Marysko, J. P. Doumerc, J. C. Grenier, J. Etourneau, "Influence of the structure on electric and magnetic properties of La0.8Na0.2Mn1-xCoxO3 perovskites", *Journal of Solid State Chemistry* **177**, 4564 (2004).

[11]   R. Robert, S. Romer, S. Hebert, A. Maignan, A. Reller, A. Weidenkaff, "Oxide materials development for solar thermoelectric power generators", *Proc. of the 2nd European Conference on Thermoelectrics* 194 (2004).

[12]   I. Matsubara, R. Funahashi, T. Takeuchi, S. Sodeoka, T. Shimizu, K. Ueno, "Fabrication of an all-oxide thermoelectric power generator", *Applied Physics Letters* **78**, 3627 (2001).

[13]   A. Weidenkaff, "Preparation and application of nanoscopic perovskite phases", *Advanced Engineering Materials* **6**, 709 (2004).

[14]   G. Kotliar, D. Vollhardt, "Strongly correlated materials: Insights from Dynamical Mean-Field Theory", *Physics Today* 53 (2004).

[15]   P. M. Raccah, J. B. Goodenough, "First-Order Localized-Electron <-> Collective electron transition in LaCoO3", *Phys.Rev. B* **155**, (1967).

[16]   C. N. R. Rao, O. Parkash, P. Ganguly, "Electronic and magnetic properties of LaNi1-xCoxO3, LaCo1-xFexO3 and LaNi1-xFexO3", *Journal of Solid State Chemistry* **15**, 186 (1975).

[17]   W. Koshibae, S. Maekawa, "Effect of spin and orbital on thermopower in strongly correlated electron systems", *Journal of Magnetism and Magnetic Materials* **258-259**, 216 (2003).

[18]   M. A. Senaris-Rodriguez, J. B. Goodenough, "Magnetic and Transport Properties of the System La1-xSrxCoO3-d (0<x<0.50)", *J.Solid State Chemistry***118**, 323 (1995).

[19]   W. Koshibae, K. Tsutsui, S. Maekawa, "Thermopower in cobalt oxides", *Phys.Rev. B* **62**, 6869 (2000).

[20]   R. R. Heikes, R. C. Miller, R. Mazelsky, "Magnetic and electrical anomalies in LaCoO3", *Physica B: Condensed Matter* **30**, 1600 (1964).

[21]   R. Robert, L. Bocher, M. Trottmann, A. Reller, A. Weidenkaff, "Synthesis and high-temperature thermoelectric properties of Ni and Ti substituted LaCoO3", *Journal of Solid State Chemistry* **179**, 3867 (2006).

[22]   P. Migiakis, J. Androulakis, J. Giapintzakis, "Thermoelectric properties of LaNi$_{1-x}$ Co$_x$O$_3$ solid solution", *J. Appl. Phys.* **94**, 7616 (2005).

[23]   G. Chen, "Phonon heat conduction in nanostructures", *International Journal of Thermal Sciences* **39**, 471 (2000).

[24]   J. Androulakis, P. Migiakis, J. Giapintzakis, "La0.95Sr0.05CoO3: An efficient room-temperature thermoelectric oxide", *Appl. Phys. Lett.* **84**, 1099 (2004).

[25]   L. Bocher, R. Robert, M. H. Aguirre, L. Schlapbach, "Thermoelectric Perovskite-Type Oxides for Geothermal and Solar Energy Conversion", *Proceedings of the 4th European Conference on Thermoelectrics* (2006).

POWER GENERATION OF p-TYPE $Ca_3Co_4O_9$/n-TYPE $CaMnO_3$ MODULE

S. Urata[1], R. Funahashi[1,2], T. Mihara[1,2], A. Kosuga[2], S. Sodeoka[2], and T. Tanaka[2]
[1] CREST, Japan Science and Technology Agency, Honmachi, Kawaguchi, Saitama 332-0012, Japan
[2] National Institute of Advanced Industrial Science & Technology, Midorigaoka, Ikeda, Osaka 563-8577, Japan

ABSTRACT

Thermoelectric modules composed of eight pairs of p-type $Ca_{2.7}Bi_{0.3}Co_4O_9$ (Co-349) and n-type $CaMn_{0.98}Mo_{0.02}O_3$ (Mn-113) bulks were constructed using Ag electrodes and paste including powder of the n-type oxide. The former bulks were prepared by hot-pressing. On the other hand, the latter were densified using a cold isostatic pressing (CIP) technique and sintered in atmospheric pressure. Dimensions of both oxide legs were 5 mm wide and thick and 4.5 mm high. An alumina plate was used as a substrate, and there was no alumina plate on the other side of the modules. When the substrate side was heated, the module can generate up to 1.0 V and 0.17 W of open circuit voltage (V_O) and maximum power (P_{max}), respectively, at a hot-side temperature of 1273 K (furnace temperature as a heat source) and a cold side temperature of 298 K (circulated water temperature) in air. But internal resistance R_I reaches at value of 1.5 Ω which is about 6 times higher than the calculated one from resistivity of both p and n-type bulks. When the substrate side is cooled, V_O and P_{max} reach 0.7 V and 0.34 W of V_O and P_{max}, respectively, at a furnace temperature of 1273 K.

1. INTRODUCTION

In view of global energy and environmental problems, research and development have been promoted in the field of thermoelectric power generation as a means of recovering vast amounts of waste heat emitted by automobiles, factories, and similar sources. Waste heat from such sources offers a high-quality energy source equal to about 70 % of total primary energy, but is difficult to reclaim due to its source amounts being small and widely dispersed. Thermoelectric generation systems offer the only viable method of overcoming these problems by converting heat energy directly into electrical energy irrespective of source size and without the use of moving parts or production of environmentally deleterious wastes. The requirements placed on materials needed for this task, however, are not easily satisfied. Not only must they possess high conversion efficiency, but must also be composed of non-toxic and abundantly available elements having high chemical stability in air even at temperatures of 800-1000 K. Thermoelectric modules are composed of intermetallic compounds, such as Bi_2Te_3, Pb-Te, and Si-Ge. Practical applications of materials like these have, however, been delayed by problems such as their low melting or decomposition temperatures, their content of harmful or scarce elements, and their cost. Recently, oxide compounds have attracted attention as promising thermoelectric materials because of their potential to overcome the above-mentioned problems [1-7]. Recently, fabrication and power generation of thermoelectric modules consisting of p-type $Ca_3Co_4O_9$ (Co-349) and n-type $LaNiO_3$ (Ni-113) legs have been reported [8]. Actually, power charging a portable phone has been successful using this module, however, thermoelectric conversion efficiency (η) was as low as 1.4 % [9]. Improvement of thermoelectric properties of Co-349 bulks and new n-type materials possessing high thermoelectric figure of merit (Z) are indispensable to enhance η. In order to solve the former issue, alignment of Co-349 grains is effective because of reduction of resistivity [10]. On the other hand, exploring new n-type oxides possessing high thermoelectric performance is carrying out eagerly. Although thermoelectric properties of materials composing the modules should be enhanced, high chemical and mechanical durability of the materials and contact resistance and strength at the junctions are also very important in practical use of the modules. The oxide materials have high chemical stability at the high temperature in air. But it is insufficient to consider the durability of modules. In this paper, after consideration of electrical and mechanical properties of oxide bulks and junctions, fabrication, power generation characteristics, and thermal durability of thermoelectric modules composed of eight pairs of oxide legs will be discussed.

2. JUNCTIONS
2.1. Preparation of bulk materials and unicouples

Modules fabricated in this study are composed of Ca$_{2.7}$Bi$_{0.3}$Co$_4$O$_9$ (Co-349) and CaMn$_{0.98}$Mo$_{0.02}$O$_3$ (Mn-113) for the p-type and n-type legs, respectively. The Co-349 powder was prepared by solid-state reaction at 1123 K for 10 h in air. As starting materials, CaCO$_3$, Co$_3$O$_4$ and Bi$_2$O$_3$ powders were used and mixed thoroughly in the stoichiometric composition. Co-349 bulks were prepared using a hot-pressing technique. The obtained Co-349 powder was hot-pressed for 20 h in air under a uniaxial pressure about 10 MPa at 1123 K to make density and grain alignment high. Preparation of the Mn-113 was started using CaCO$_3$, Mn$_2$O$_3$ and MoO$_3$ powders. These powders were mixed well and treated at 1273 K for 12 h in air to prepare the Mn-113 powder. The powder was densified using a cold isostatic pressing (CIP) technique for 3 h under about 150 MPa. After the CIP process, the precursor pellets were sintered at 1473 K 12 h in air. Mn-113 bulks can be densified well by CIP and sintering under the atmospheric pressure even without hot-pressing. Both bulks were cut to provide leg elements with a cross-sectional area of 5.0 × 5.0 mm^2 and length of 4.5 mm.

An alumina plate possessing dimensions of 5.0 mm wide, 10.0 mm long, and 1.0 mm thick was used as a substrate. An Ag sheet with 5.0 mm in width, 10.0 mm in length, and 50 μm in thickness was attached on the alumina plate using Ag paste to achieve electrical conduction. The oxide legs and Ag sheet on the alumina plate were adhered by Ag paste mixed with n-type oxide powder. Compositions of the oxide powder used were identical to those of the n-legs. The same precursor powders were pulverized by ball milling to obtain a grain size smaller than 10 μm, then mixed in varying ratios with the Ag paste (0, 1.5, 3, 6, or 10 wt.%). After connection of the legs and substrate, the wet Ag paste was dried at 373 K and solidified by heating at 1123 K under a uniaxial pressure of 6.4 MPa in air. The unicouple is shown in Fig. 1.

Figure 1. Thermoelectric unicouple composed of Co-349 and Mn-113 bulks.

2.2. Measurement of internal resistance R_I

    Measurement of internal resistance (R_I) was carried out using a standard DC four terminal method. The current terminals for the substrate were attached on the opposite side. The voltage terminals were prepared on the vertical sides of both p- and n-legs just under the current terminals.

2.3. Results and discussion

    Temperature dependence of R_I for Co-349/Mn-113 unicouples prepared by Ag paste including various ratios of Mn-113 powder is shown in Fig. 2. It is clear that mixing the Mn-113 powder into the Ag paste is effective to reduce R_I. This is due to the reduction of contact resistance (R_C) between the oxide legs and the Ag electrodes. An increase in R_I, however, is observed when the Ag paste containing 6 and 10 wt% of oxide powders is used. The optimum-mixing ratio of the oxide powders is present around 1.5-3 wt%.

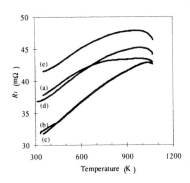

Figure 2. Temperature dependence of R_I for Co-349/Mn-113 unicouples prepared using Ag paste including Mn-113 powder. 0 wt% (a), 1.5 wt% (b), 3 wt% (c), 6 wt% (d) and 10 wt% (e) of oxide powder was incorporated.

3. MECHANICAL PROPERTIES

3.1. Sample preparation and measurement

    The Co-349 and Mn-113 bulk samples for the investigation of mechanical properties were prepared in the same conditions as mentioned above. The bulks were cut into dimensions of 4.0 mm wide, 3.0 mm thick, and 40.0mm long for three-point bend test and 5.0 mm wide, 5.0 mm thick, and 10.0 mm long for thermal expansion coefficient.

    Three-point bend test was carried out in room temperature, the loading speed was 0.5 mm/min and span length was 30.0mm (Fig. 3). In the case of Co-349 bulks, measurement was performed in the loading direction perpendicular and parallel to the hot-pressing axis. Linear thermal expansion coefficient was measured using a differential dilatometer at 323-1073 K in air.

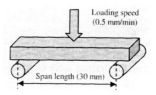

Figure 3. Geometry of the three-point bend test.

3.2. Results and discussion

As mentioned below, only the Mn-113 legs were broken in the modules after power generation. To investigate the cause of the destruction, thermal expansion coefficient and three-point bending strength were investigated. Thermal expansion coefficient for the Co-349 bulks is lower than that for Mn-113 bulks and closer to the alumina plates for the substrate of the modules (Fig. 4).

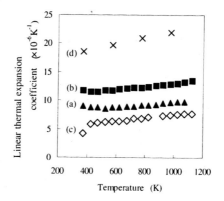

Figure 4. Temperature dependence of linear thermal expansion coefficient of (a) Co-349, (b) Mn-113 bulks, (c) alumina plate for the substrate of the modules, and (d) and Ag ingot.

The three-point bending strength of the Co-349, Mn-113 bulks, and alumina plates are shown in Fig. 5. The alumina substrates show the highest bending strength. Anisotropy of bending strength is observed in the Co-349 bulks. The strength in the case of loading direction parallel to the hot-pressing axis is higher than perpendicular to the hot-pressing axis. The bending strength of the Mn-113 bulks is lower than that of the Co-349 bulks in both cases of the loading direction parallel and perpendicular to the hot-pressing axis. The differences in thermal expansion coefficient and bending strength between the Co-349 and Mn-113 bulks lead to the destruction of the latter legs only as mentioned below.

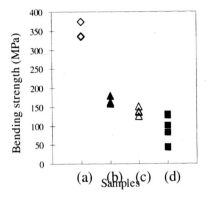

Figure 5. Three-point bending strength of (a) alumina plates, (b) Co-349 (loading direction parallel to hot-pressing axis), (c) Co-349 (loading direction perpendicular to hop-pressing axis), (d) Mn-113 bulks.

4. THERMOELECTRIC MODULES

4.1. Fabrication

The oxide legs were prepared as shown in 2.1. The dimensions of the legs were the cross-sectional area of 5.0×5.0 mm^2 and length of 4.5 mm. Electrodes were formed on one side of surface of an alumina plate (36.0mm × 34.0mm × 1.0mm thick) using Ag paste including oxide powder and Ag sheets with a thickness of 50 μm. An alumina plate was used for one side of the module as a substrate. On the other hand, no substrate was used for the other side. This structure is effective to prevent the contact between oxide legs and Ag electrodes from peeling by the deformation of the module. For connection between the oxide legs and electrodes, Ag paste including 3 wt% of n-type powder was used as an adhesive paste because of low R_l as shown in Fig. 2. This oxide/Ag composed paste was applied by screen printing on the Ag electrodes. The eight pairs of p and n-type oxide legs were put on them alternatively. The precursor module was solidified at 1123 K under a uniaxial pressure of 6.4 MPa for 3 h in air. The perfect Co-349/Mn-113 module is shown in Fig. 6.

Figure 6. A thermoelectric module composed of eight pairs of Co-349/Mn-113 legs.

4.2. Evaluation of thermoelectric power generation and durability

The module was put on a plate shape furnace to heat at 673-1273 K (hot-side temperature T_H). A cooling jacket was put on the other side the module and cooled at 298 K by water circulation (T_C). Measurement of the power generation, in which current (I)-voltage (E) lines and I-power (P) curves, was carried out in air by changing load resistance using an electronic load system. Internal resistance R_I of the module corresponded to the slope of the I–E lines. The power generation performances were compared when the substrate side was heated and cooled.

Durability against heating-cooling cycles was investigated for the 15 pieces of modules. The electrical furnace temperature of T_H was set at 523 K and on the other side of the modules was cooled by water circulation at 293 K. After heating for 7 h with continuous power generation, the modules were cooled down to room temperature. This heating-cooling cycle was carried out 4-times. R_I of the modules was measured using a standard DC four terminal method before and after the heating-cooling cycles.

4.3. Results and discussion

The module can generate up to 1.0 V and 0.17 W of V_O and P_{max}, respectively at the furnace temperature of 1273 K (T_H) when the substrate side is heated (Fig. 7). On the other hand, V_O and P_{max} of the module reach 0.7 V and 0.34 W, respectively when the substrate side is cooled as shown in Fig. 7. P_{max} measured by cooling the substrate is higher by twice than that measured by heating the substrate. Temperature differential in the module is larger by cooling no-substrate side than the substrate side. The destruction of the oxide legs, however, happens and enhance R_I in the module heated the substrate.

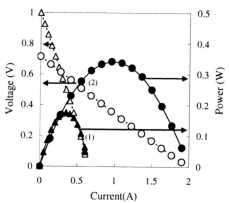

Figure 7. Power generation characters of eight pairs Co-349/Mn-113 module when the furnace temperature (T_H) was 1273K. (1) Substrate side was heated and (2) substrate side was cooled.

The R_I value calculated from resistivity of both p-type and n-type bulks at T_H of 1273 K and T_C of 298 K is as low as 0.250 Ω and lower than the measured one of the virgin modules [11] (Fig. 8). R_I tends to increase with increasing T_H. Especially, high R_I is remarkable in the case of heating the substrate. As mentioned above, the differential of thermal expansion coefficient between the Mn-113 leg, which obtain low mechanical strength, and the alumina plate is large. Figure 9 shows the destruction of the module. All destruction happened in the n-type legs. A reason of the destruction of the Mn-113 legs is the differential of thermal expansion coefficient between the alumina substrate. All broken points are not the junctions but in horizontal direction at about 1 mm height of the Mn-113 legs from the surface of the alumina substrate. Therefore, the destruction of the Mn-113 legs leads to the remarkable enhancement in R_I by heating the substrate. The mechanical strength of the Mn-113 bulk is necessary to be improved. Higher R_I is due to not only destruction but also high R_C between oxide legs and Ag electrodes. In order to reduce R_I, R_C especially between the Ag electrodes and the n-type legs should be suppressed by improvement of the adhesive materials.

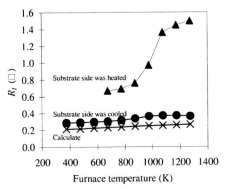

Figure 8. Measured and calculated R_I of the eight pairs of Co-349/Mn-113 modules.

Figure 9. Destruction of the Mn-113 legs in the module after power generation.

Durability against heating-cooling cycles of the 15 peaces of modules was evaluated. R_I was increased after the cycles (Fig. 10). Because 5 modules (module number 4, 7, 8, 14 and 15) were broken after the cycles completely, R_I could not be measured. Considering high R_I even at lower T_H than 523 K, the destruction of the n-type legs seems to start before T_H reaching at 523 K and extend with increasing temperature.

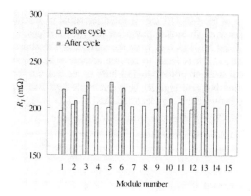

Figure 10. R_I of the modules before and after the heating-cooling cycles. R_I for the module number 4, 7, 8, 14 and 15 could not be measured after the cycles because of complete destruction.

5. CONCLUSION

Thermoelectric modules consisting of eight pairs of p-type Co-349 and n-type Mn-113 legs have been fabricated. It is clear that the incorporation of the n-type oxide powder into Ag paste by 1.5-3 wt% is effective to reduce R_C. Thermal expansion coefficient of the Co-349 bulk is closer than that of the Mn-113 one to the alumina substrate. And the three-point bending strength of the Co-349 bulk is higher than that of Mn-113 one. These are the reasons of the destruction of the Mn-113 legs near the alumina substrate in the modules by heating.

The module can generate up to V_o and P_{max} 1.0 V and 0.17 W, respectively by heating the substrate side at the furnace temperature of 1273 K and cooling the opposite side at the circulated water temperature of 298 K. On the other hand, V_O and P_{max} are as high as 0.7 V and 0.34 W, respectively by cooling the substrate side. R_I of the modules was increased after heating for power generation. This is due to the destruction of the n-type legs. The Mn-113 legs are damaged by thermal stress between the alumina substrate. The module has a potential of 2.0 % of conversion calculated by ZT (dimension-less figure of merit) values at T_H of 1073 K and temperature differential of 600 K, if the mechanical properties and R_C are improved.

REFERENCES

[1] I. Terasaki et al., "Large thermoelectric power in $NaCo_2O_4$ single crystals", *Phys. Rev. B*, **56** 12685-87 (1997).

[2]. R. Funahashi et al., "An Oxide Single Crystal with High Thermoelectric Performance in Air", *Jpn. J. Appl. Phys.*, **39** L1127-29 (2000).

[3] Y. Miyazaki et al., "Low-Temperature Thermoelectric Properties of the Composite Crystal $[Ca_2CoO_{3.34}]_{0.614}[CoO_2]$", *Jpn. J. Appl. Phys.*, **39** L531-33 (2000).

[4] R. Funahashi et al. "Thermoelectric properties of Pb- and Ca-doped $(Bi_2Sr_2O_4)_xCoO_2$ whiskers", *Appl. Phys. Lett.*, **79** 362-64 (2001).

[5] M. Ohtaki et al., "Electrical Transport Properties and High-Temperature Thermoelectric Performance of $(Ca_{0.9}M_{0.1})MnO_3$ (M = Y, La, Ce, Sm, In, Sn, Sb, Pb, Bi)", *J. Solid State Chem.*, **120** 105-11 (1995).

[6] Y. Masuda et al., "Structure and Thermoelectric Transport Properties of Isoelectronically Substituted $(ZnO)_5In_2O_3$", *J. Solid State Chem.*, **150** 221-27 (2000),

[7] W. Shin et al., "Li-Doped Nickel Oxide as a Thermoelectric Material", *Jpn. J. Appl. Phys.*, **38** L1336-38 (1999).

[8] R. Funahashi et al., "$Ca_{2.7}Bi_{0.3}Co_4O_9$/$La_{0.9}Bi_{0.1}NiO_3$ thermoelectric devices with high output power density", *Appl. Phys. Lett.*, **85**, 1036-1038 (2004).

[9] R. Funahashi et al., "A portable thermoelectric-power-generating module composed of oxide devices", *J. Appl. Phys.*, **99**, 066117-19 (2006).

[10] E. Guilmeau et al., "Thermoelectric properties–texture relationship in highly oriented $Ca_3Co_4O_9$ composites", *Appl. Phys. Lett.*, **85**, 1490-92 (2004).

[11] T. Noguchi et al., The Papers of Technical Meeting on Frontier Technology and Engineering, IEE Japan, **7** (2003) in Japanese.

THERMOELECTRIC PROPERTIES OF Pb AND Sr DOPED Ca$_3$Co$_4$O$_9$

Hiroshi Nakatsugawa[1], Hyeon Mook Jeong[1], Natsuko Gomi[1] and Hiroshi Fukutomi[1]
[1]Graduate school of Engineering, Yokohama National University
79-5 Tokiwadai, Hodogaya-ku
Yokohama 240-8501, Japan

Rak Hee Kim[2]
[2]School of Advanced material Engineering, Changwon National University
9 Sarim-dong, Changwon
Gyeongnam 641-773, Korea

ABSTRACT

We have prepared polycrystalline specimens of $[(Ca_{1-x}Pb_x)_2CoO_3]_{0.62}CoO_2$ and $[(Ca_{1-y}Sr_y)_2CoO_3]_{0.62}CoO_2$ using the conventional solid-state reaction method, and investigated the Pb substitution effect on the thermoelectric and magnetic properties. With the Pb substitution, both the electrical resistivity and Seebeck coefficient do not change drastically. This is attributed to the carrier concentration of samples. Seebeck and Hall coefficient measurements reveal that the major charge carriers in the samples are holes, however, the carrier concentration does not change with increasing x. The neutron powder diffraction technique and the magnetic susceptibility measurements also reveal that Pb ions take divalent state in the rock salt type Ca$_2$CoO$_3$ block layer. The average valence state of Co ions in the CdI$_2$ type CoO$_2$ sheet was 3.1+ and that of Co ions in the block layer was 3.6+. The resulting dimensionless figure of merit for the x = 0.02 sample at room temperature becomes 0.024, which is approximately equal to the corresponding values of a polycrystalline sample of NaCo$_2$O$_4$.

INTRODUCTION

Since the discovery of large thermoelectric power in the layered compounds NaCo$_2$O$_4$ and Ca$_3$Co$_4$O$_9$,[1-4] misfit-layered cobalt oxides particularly have attracted much interest as candidates for thermoelectric (TE) materials. Figure 1 shows the initial structure model projected in perspective from b-axis (left) and from a-axis (right). As shown in Fig.1, the Crystal structure of Ca$_3$Co$_4$O$_9$ consists of an alternate stack of a distorted three-layered rock salt (RS)-type Ca$_2$CoO$_3$ block layer (BL) and a CdI$_2$-type CoO$_2$ conducting sheet parallel to the c-axis.[3, 5, 6] The CdI$_2$-type [CoO$_2$] subsystem and the RS-type [Ca$_2$CoO$_3$] BL subsystem have common a- and c-axes and beta angles. Owing to the size difference between the RS-type BL and the CoO$_2$ sheet, however, the compound has an incommensurate periodicity parallel to the b-axis. The resulting structural formula becomes $[Ca_2CoO_3]_pCoO_2$, where p equals $b_{CoO_2}/b_{Ca_2CoO_3} = 0.62$ is an oxygen nonstoichiometry. The chemical formula can be approximately represented as Ca$_{1.24}$Co$_{1.62}$O$_{3.86}$. Throughout this paper, we will use the structural formula of the compound instead of the chemical formula. Such an anisotropic structure of the compound is believed to be favorable in realizing large absolute value of Seebeck coefficient, S, and low thermal conductivity, κ, necessary for good TE compounds. Furthermore, the RS-type BL can be regarded as a charge reservoir, which introduces hole carriers into the CoO$_2$ sheet. In addition, a CoO$_2$ triangular

171

lattice in the CoO$_2$ sheet should play an important role for realizing low electrical resistivity, ρ, namely, large TE power factor, S^2/ρ.

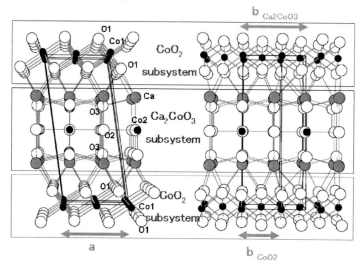

Fig. 1 Initial structure model of [Ca$_2$CoO$_3$]$_\rho$CoO$_2$.

As is well known, the $3d$ orbital of an octahedrally coordinated Co ion splits into doubly degenerated upper e_g and triply degenerated lower t_{2g} levels. The t_{2g} levels further split into another doubly degenerated $e_g{'}$ levels and a non-degenerated a_{1g} level due to the rhombohedral distortion of the octahedron. Since ^{59}Co NMR studies of metallic NaCo$_2$O$_4$ confirmed that both Co^{3+} and Co^{4+} ions are in the low-spin (LS) states,[7] the electronic configuration of each ion is Co^{3+}: $\left(e_g{'}\right)^4 \left(a_{1g}\right)^2$ and Co^{4+}: $\left(e_g{'}\right)^4 \left(a_{1g}\right)^1$, respectively. According to the band calculation of NaCo$_2$O$_4$,[8] the density of states (DOS) near the Fermi level (E$_F$) consists of the narrow (localized) a_{1g} band and the broad (itinerant) $e_g{'}$ band. The height of E$_F$ depends on the nominal valence state of the Co ions. For the case of Co^{3+}, E$_F$ is located at the upper edge of the a_{1g} band, and therefore, high ρ is expected. With increasing hole carriers, the nominal valence state of the Co ions gradually close in Co^{4+} and E$_F$ crosses the bands, where a_{1g} and $e_g{'}$ are hybridized. Thus, low ρ is then expected for such a mixed valent compound.

The polycrystalline Ca$_3$Co$_4$O$_9$ sample typically exhibits $S = 130$ $\mu V/K$, $\rho = 1.5 \times 10^{-4}$ Ωm and $\kappa = 1.0$ W/mK at room temperature.[4] For practical use, an appreciable decrease in ρ must be achieved because $\rho = 1.5 \times 10^{-4}$ Ωm at room temperature is about one order of magnitude higher than that of Bi$_2$Te$_3$-based TE materials. Much effort has been devoted to decrease ρ while maintaining a large S and low κ, through partial substitutions for Ca atoms in the RS-type BL. The optimization of the valence state of Co ions in the RS-type BL is a key issue to maximize the TE properties because ρ and S are highly dependent on the nominal valence state of Co ions in the CoO_2 sheet. In fact, Li et al.[9] and Funahashi et al.[10] reported a marked increase in TE performance by partial substitution of Bi for Ca. However, no studies have ever tried to report the effect of Pb-substitution for Ca. Thus, we have employed a high-resolution neutron powder diffraction technique to investigate the modulated crystal structure of the Pb and Sr doped $[Ca_2CoO_3]_{0.62}CoO_2$ polycrystalline samples. This paper is intended as an investigation of the neutron powder diffraction technique and also measurements of electrical resistivity, Seebeck coefficient, Hall coefficient, thermal conductivity and magnetic susceptibility to clarify both the TE properties and the valence state of Co ions.

EXPERIMENT

The polycrystalline specimens of $[(Ca_{1-x}Pb_x)_2CoO_3]_{0.62}CoO_2$ and $[(Ca_{1-y}Sr_y)_2CoO_3]_{0.62}CoO_2$ were prepared by the conventional solid-state reaction method starting from powder mixture of $CaCO_3$ (99.9 %), $SrCO_3$ (99.9%), PbO (99.9 %) and Co_3O_4 (99.9 %) with a stoichiometric cation ratio. After calcination in air at 920 □ for 12 h, the calcined powders were pressed into pellets and sintered in pure flowing oxygen gas at 920 □ for 24 h. The obtained well-crystallized single-phase samples were annealed in pure flowing oxygen gas at 700 □ for 12 h and then quenched into distilled water to control oxygen nonstoichiometry.[11] Through the synthesis, the nominal b-axis ratio $b_{CoO_2}/b_{Ca_2CoO_3}$ was fixed at about 0.62 in all samples.

Neutron powder diffraction (ND) data were collected at room temperature using the Kinken powder diffractometer for high efficiency and high resolution measurements (HERMES) of Institute for Materials Research (IMR), Tohoku University, installed at the JRR-3M reactor in Japan Atomic Energy Research Institute (JAERI), Tokai.[12] Neutrons with a wavelength of 0.18265 nm were obtained by the 331 reflection of the Ge monochromator. The ND data were collected on thoroughly ground powders by a multiscanning mode in the 2θ range from 3.0°to 153.9°with a step width of 0.1°and were analyzed using a Rietveld refinement program, PREMOS 91,[13] adopting a superspace group of $C2/m(1p0)s0$, where the CdI_2 - type $[CoO_2]$ subsystem has $C2/m$ symmetry while the RS – type BL $[Ca_2CoO_3]$ subsystem has $C2_1/m$ symmetry. The crystal structures and interatomic distance plots were obtained with the use of PRJMS and MODPLT routines, respectively; both were implemented in the PREMOS 91 package.

Measurements of electrical resistivity, Seebeck coefficient and Hall coefficient were carried out in temperature range from 80 K to 385 K. The electrical resistivity, ρ, was measured by the van der Pauw technique with a current of 10 mA in He atmosphere. The Seebeck coefficient, S, measurement was carried out on a sample placed between two blocks of oxygen-free high conductivity (OFHC) silver. The thermocouples were welded to the reverse sides of the OFHC silver to measure temperature difference $\Delta T = 3$ K and thermoelectric power, ΔV. The slope ($d\Delta V/d\Delta T$) obtained by the least-squares method yields the Seebeck coefficient, S, at each measurement temperature. The Hall coefficient, R_H, measurement was performed on flat square pieces of materials with a current of 100 mA in a magnetic field of 8500 Oe by the van der Pauw technique. The Hall carrier concentration n was determined from R_H using $n = 1/eR_H$, where e is the electron charge, assuming a scattering factor equal to 1 and a single carrier model. Furthermore, the Hall mobility, μ, was determined from ρ and R_H using $\mu = R_H/\rho$.

Thermal conductivity was measured at room temperature and magnetic susceptibility measurements were carried out in the temperature range from 2 K to 350 K. The thermal conductivity, κ, was calculated from $\kappa = A \cdot C \cdot D$, where A is the thermal diffusivity, C is the specific heat and D is the measured sample density. The thermal diffusivity was measured using the laser flash method (ULVAC, TC-3000) at room temperature. The magnetic susceptibility, χ, was measured using a superconducting quantum interference device (SQUID) magnetometer (Quantum Design, MPMS) under the zero-field cooling (ZFC) and field cooling (FC) conditions in a magnetic field of 10 Oe.

RESULTS AND DISCUSSION

Figure 2 shows the temperature dependence of the electrical resistivity of the specimens with [(Ca$_{1-x}$Pb$_x$)$_2$CoO$_3$]$_{0.62}$CoO$_2$ and [(Ca$_{1-y}$Sr$_y$)$_2$CoO$_3$]$_{0.62}$CoO$_2$ measured at the temperature range from 80 K to 385 K. In particular, the sample with $x = 0.02$ shows lowest electric resistivity, 1.67×10^{-4} Ωm at room temperature. All the samples show metallic behavior down to around 90 – 100 K. With further decrease in the temperature, the samples turn to be semiconducting behavior due to a carrier localization of a ferrimagnetic transition at around 20 K and the possible occurrence of the spin-density-wave (SDW) at around 100 K. Sugiyama et al.[14] have suggested that Co^{4+} ions in the RS – type BL play an important role in including the SDW transition. Furthermore, it is reported that the ferrimagnetic transition is originated from the modulated magnetic sublattices of Co^{4+} ions in the BL subsystem.[14]

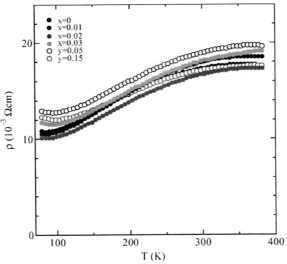

Fig. 2 Temperature dependence of the electrical resistivity ρ

The temperature dependence of the absolute value of Seebeck coefficient is shown in Fig.3. It can be seen clearly that at the temperatures above 100 K the Seebeck coefficient of $[(Ca_{1-x}Pb_x)_2CoO_3]_{0.62}CoO_2$ and $[(Ca_{1-y}Sr_y)_2CoO_3]_{0.62}CoO_2$ compounds shows weak temperature dependence. We have confirmed that all the samples have been found to show similar S. The effect of the Pb^{2+} substitution on the S values could be interpreted by the framework of the model which was proposed to explain the large Seebeck coefficient of the NaCo$_2$O$_4$. In that model, the Seebeck coefficient of high temperature limit can be estimated by using the modified Heikes formula: [15]

$$S(T \to \infty) = -\frac{k_B}{e} \ln\left[\frac{g_3}{g_4}\frac{c}{1-c}\right]$$

, where g_3, g_4 are the number of the degenerated configurations of the Co^{3+} and Co^{4+} states in the CoO$_2$ sheet while $c = $ Co^{4+}/Co is the fraction of Co^{4+} holes on the Co sites in the [CoO$_2$] subsystem. Since the electronic configuration of Co^{3+} and Co^{4+} ion is the low spin state, the number of the degenerated configurations of the Co^{3+} and Co^{4+} states in the CoO$_2$ sheet is $g_3 = 1$ and $g_4 = 6$, respectively. Thus according to the above formula, the $S(T \to \infty)$ values would be steady with no change in c value. In the present study, the substitution of divalent Pb^{2+} for divalent Ca^{2+} would keep the hole concentrations. Therefore, $[(Ca_{1-x}Pb_x)_2CoO_3]_{0.62}CoO_2$ and $[(Ca_{1-y}Sr_y)_2CoO_3]_{0.62}CoO_2$ compounds shows similar S as shown in Fig.3.

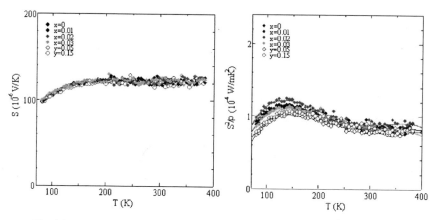

Fig. 3 Temperature dependence of S.　　**Fig. 4** Temperature dependence of S^2/ρ.

The TE power factor, S^2/ρ, of $[(Ca_{1-x}Pb_x)_2CoO_3]_{0.62}CoO_2$ and $[(Ca_{1-y}Sr_y)_2CoO_3]_{0.62}CoO_2$ as a function of temperature is represented in Fig.4. In particular, the sample with $x = 0.02$ exhibits largest TE power factor, $9.5\times10^{-5}\ W/mK^2$ at room temperature and shows a broad maximum at around $100 - 150$ K. With further decrease in temperature, the S^2/ρ value decreases gradually. Since the Seebeck coefficient $S(T)$ behavior does not change very much, as we have mentioned before, the TE power factor is primarily dependent on the value of electrical resistivity $\rho(T)$

Figure 5 shows the temperature dependence of the Hall coefficient $R_H(T)$ of the samples with $[(Ca_{1-x}Pb_x)_2CoO_3]_{0.62}CoO_2$ and $[(Ca_{1-y}Sr_y)_2CoO_3]_{0.62}CoO_2$ measured at the temperature range from 80 K to 385 K. All the samples show a positive value of R_H and similar values. Hall coefficient measurements reveal that the major charge carriers in all the samples are holes. Moreover, the signs of Hall coefficient were consistent with those of Seebeck coefficient. Hall carrier concentration $n = 1/eR_H$ and Hall mobility $\mu = R_H/\rho$ at room temperature are also shown in insets.

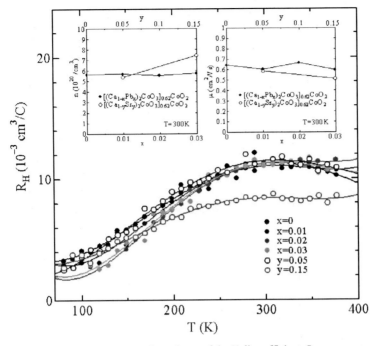

Fig. 5 Temperature dependence of the Hall coefficient R_H

We assume that the Pb atoms can only substitute for Ca. site. Figure 6 shows the observed, calculated and difference intensities of the HERMES data for $x = 0.02$ sample. The final R_{wp} factor was 8.0 % and the lattice parameters were refined to $a = 0.48384(3)$ nm, $b_{CoO_2} = 0.28248(1)$ nm, $c = 1.08865(6)$ nm and $\beta = 98.19(1)°$ for $[CoO_2]$ subsystem and $b_{RS} = 0.45786(7)$ nm for RS-type subsystem. The resulting $p = b_{CoO_2}/b_{Ca_2CoO_3} = 0.6169(5)$ corresponds to the stoichiometry of the $x = 0.02$ sample, i.e., $[(Ca_{0.98}Pb_{0.02})_2CoO_3]_{0.6169}CoO_2$.

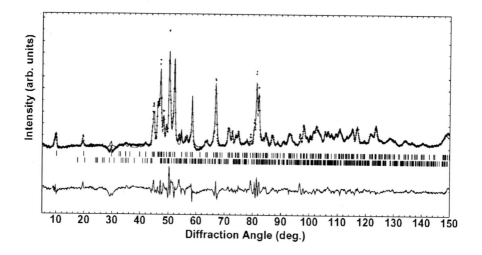

Fig. 6 Observed, calculated and difference intensities of powder neutron diffraction data (HERMES data) for $x = 0.02$. Short vertical lines below the intensities indicate the peak positions of the main (upper) and satellite (lower) reflections for the CdI_2-type $[CoO_2]$ sheet and RS-type $[Ca_2CoO_3]$ BL subsystem. The differences between the observed and calculated patterns are shown below the vertical lines.

The variation of the atomic positional modulation can be further understood by the plot against t', a complementary coordinate in the (3+1)-dimensional superspace.[13] In the present samples, t' is defined as $-px_2 + x_4 = -0.62x_2 + x_4$. Figure 7 shows Co1 – O (upper panels) distances in the $[CoO_2]$ subsystem and Co2 – O (lower panels) distances in the RS-type subsystem of (a) $x = 0$ (left) and (b) $x = 0.02$ (right) plotted against t'. Co2 site at $z = 1/2$ has six oxygen neighbors, with four equatorial O2 and two apical O3 atoms. Among these bonds, two apical Co2 – O3 bonds are fairly shorter than the other four equatorial Co2 – O2 bonds, ranging from 0.17 to 0.195 nm for $x = 0$ and from 0.165 to 0.19 nm for $x = 0.02$. The mean distance of Co2 – O3 bonds for $x = 0$ does not change very much as that of Co2 – O3 bonds for $x = 0.02$, i.e., both distances are 0.18 nm. This fact indicates that the Co ions in the RS-type subsystem are not replaced by tetravalent lead ions in the x range from 0 to 0.02. On the other hand, the four equatorial Co2 – O2 bonds for $x = 0$ (lower left panel) show small modulation amplitudes from 0.205 and 0.275 nm, relative to the $x = 0.02$ (lower right panel) of 0.18 and 0.3 nm. However, the mean distances of the Co2 – O2 bonds of the two phases are almost equal, i.e., 0.24 nm. In contrast to the Co2 – O bonds, the six Co1 – O1 bonds illustrated in the upper panels remain stable with increasing x, where the mean distance of the Co1 – O1 bonds is 0.19 nm.

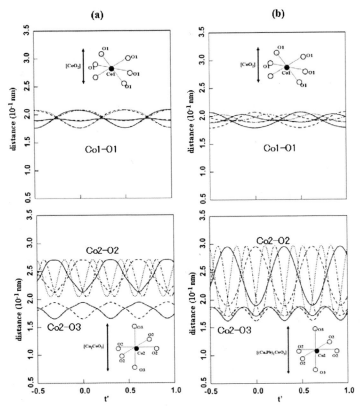

Fig. 7 Six Co1 – O1 distances (top left and right), four Co2 – O2 distances (bottom left and right) and two Co2 – O3 distances (bottom left and right) against a complementary coordinate, $t' = -px_2 + x_4 \, \square \, -0.62x_2 + x_4$, in superspace for (a) x = 0 (left) and (b) x = 0.02 (right).

Relative to the Co – O distances, the (Ca, Pb) – O distances exhibit markedly change, due to the Pb substitution as demonstrated in Figs. 8(a) and 8(b), respectively. As illustrated in Fig.8, (Ca, Pb) site at $z = 0.72 \, \square \, 0.73$ are coordinated to an apical O2 atom and four equatorial O3 atoms in the RS-type subsystem (see lower panels) and three O1 atoms in the [CoO₂] subsystem (see upper panels). The (Ca, Pb) – O1 distances vary as shown in the upper panels. In contrast, the (Ca, Pb) – O2 and (Ca, Pb) – O3 distances vary as illustrated in the lower panels. Since the O2 and O3 atoms belong to the same RS-type subsystem, the Ca – O2 and Ca – O3 distances in x = 0 are moderately altered between 0.23 and 0.26 nm (lower left panel) with a mean distance of 0.245 nm, which is comparable to the typical ionic Ca – O bond lengths. The (Ca, Pb) – O2 and (Ca, Pb) – O3 distances for x = 0.02 shift to a considerably longer range from 0.215 and 0.283 nm (lower right panel), then the mean distance of five bonds was found to change markedly

(2.49 nm). This fact indicates that the divalent Ca^{2+} ions are partially replaced by the divalent Pb^{2+} ions with increasing x.

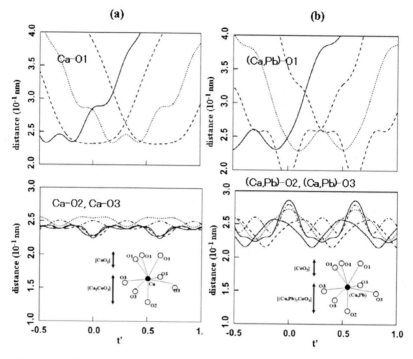

Fig. 8 Three Ca – O1 distances (top left), a Ca – O2 distance (bottom left) and four Ca – O3 distances (bottom left) against a complementary coordinate, $t' = -px_2 + x_4 = -0.62x_2 + x_4$, in superspace for (a) x = 0 (left). Three (Ca, Pb) – O1 distances (top right), a (Ca, Pb) – O2 distance (bottom right) and four (Ca, Pb) – O3 distances (bottom right) against a complementary coordinate t' in superspace for (b) x = 0.02 (right).

Let us evaluate the valence state of Co ions in $[(Ca_{1-x}Pb_x)_2CoO_3]_{0.62}CoO_2$ ($0 \square x \square 0.03$) assuming that the valence state of substituted lead ions are mainly divalent Pb^{2+}. Sugiyama et al.[14] have suggested that the polycrystalline Ca₃Co₄O₉ sample exhibits ferrimagnetic transition at $T_C = 19\ K$, due to an antiferromagnetic (AF) order of the Co1 in the $[CoO_2]$ subsystem and Co2 in the RS – type subsystem. Figures 9(a) – (d) represent the temperature dependence of the inverse-magnetic susceptibility, $(\chi - \chi_0)^{-1}$, where red line shows appropriately fitting the $(\chi - \chi_0)^{-1}$ data. $\chi_0 = 7 \times 10^{-4}\ emu/mol(Co)$ is a temperature-independent term and $T_C = 19\ K$ is a ferrimagnetic transition temperature. The insets show temperature dependence of magnetic susceptibility χ in a magnetic field of 10 Oe under ZFC (open circles) and FC (solid circles).

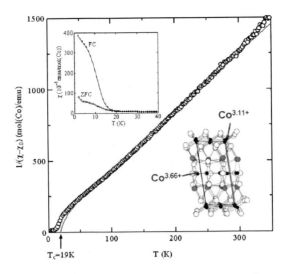

Fig. 9(a) Temperature dependence of inverse magnetic susceptibility $(\chi - \chi_0)^{-1}$ for $x = 0$.

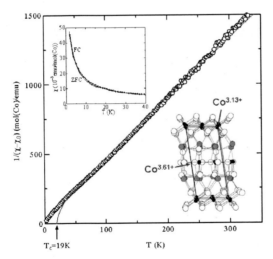

Fig. 9(b) Temperature dependence of inverse magnetic susceptibility $(\chi - \chi_0)^{-1}$ for $x = 0.01$.

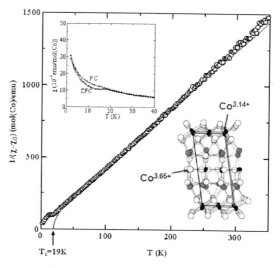

Fig. 9(c) Temperature dependence of inverse magnetic susceptibility $\left(\chi-\chi_0\right)^{-1}$ for $x = 0.02$.

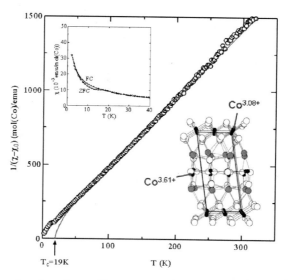

Fig. 9(d) Temperature dependence of inverse magnetic susceptibility $\left(\chi-\chi_0\right)^{-1}$ for $x = 0.03$.

CONCLUSION

The polycrystalline specimens of have been prepared using the conventional solid-state reaction method, and the effect of Pb substitution on the TE properties have been studied. The sample with $x = 0.02$ shows rather low ρ and large S suggesting promising TE materials. Seebeck coefficient shows similar large S in all the samples because the substitution of Pb^{2+} for Ca^{2+} would keep the carrier (holes) concentration. In fact, the Co2 - O3 modulation reveal that Pb ions take mainly Pb^{2+} in the RS-type BL. The magnetic susceptibility measurements also show that Co1 and Co2 sites take the average valence state of Co$^{3.1+}$ and Co$^{3.6+}$, respectively. This suggests that Pb ions can hardly take Pb^{4+} in the RS-type BL of all samples.

ACKNOWLEDGMENTS

The Hall effect measurement system in Instrumental Analysis Center and the SQUID magnetometer in Ecotechnology System Laboratory, Yokohama National University, were used. The thermal conductivity measurement was carried out in AGNE Technical Center. This study was partly supported by the IWATANI NAOJI foundation and the IKETANI foundation for promotion of science and engineering.

REFERENCES

[1] I. Terasaki, Y. Sasago and K. Uchinokura: Phys. Rev. B **56** (1997) R12685.
[2] S. Li, R. Funahashi, I. Matsubara, K. Ueda and H. Yamada: J. Mater. Chem. **9** (1999) 1659.
[3] A. C. Masset, C. Michel, A. Maignan, M. Hervieu, O. Toulemomde, F. Aruder, B. Raveau and J. Hejtmanek: Phys. Rev. B **62** (2000) 166.
[4] Y. Miyazaki, K. Kudo, M. Akoshima, Y. Ono, Y. Koike and T. Kajitani: Jpn. J. Appl. Phys. **39** (2000) L531.
[5] S. Lambert, H. Leligny and D. Grebille: J. Solid State Chem. **160** (2001) 322.
[6] Y. Miyazaki, M. Onoda, T. Oku, M. Kikuchi, Y. Ishii, Y. Ono, Y. Morii and T. Kajitani: J. Phys. Soc. Jpn. **71** (2002) 491.
[7] R. Ray, A. Ghoshray, K. Ghoshray and S. Nakamura: Phys. Rev. B **59** (1999) 9454.
[8] D. J. Singh: Phys. Rev. B **61** (2000) 13397.
[9] S. Li, R. Funahashi, I. Matsubara, K. Ueno, S. Sodeoka and H. Yamada: Chem. Mater. **12** (2000) 2424.
[10] R. Funahashi, I. Matsubara, H. Ikuta, T. Takeuchi, U. Mizutani and S. Sodeoka: Jpn. J. Appl. Phys. **39** (2000) L1127.
[11] J. Shimoyama, S. Horii, K. Otzschi, M. Sano and K. Kishio: Jpn. J. Appl. Phys. **42** (2003) L194.
[12] K. Ohoyama, T. Kanouchi, K. Nemoto, M. Ohashi, T. Kajitani and Y. Yamaguchi: Jpn. J. Appl. Phys. **37** (1998) 3319.
[13] A.Yamamoto: Acta Cryst. A **49** (1993) 831.
[14] J. Sugiyama, H. Itahara, T. Tani, J. H. Brewer and E. J. Ansaldo: Phys. Rev. B **66** (2002) 134413.
[15] W. Koshibae, K. Tsutsui and S. Maekawa: Phys. Rev. B **62** (2000) 6869.

THERMOELECTRIC PROPERTIES OF MIX-CRYSTAL COMPOUND Mg_2Si-Mg_3Sb_2

Dongli Wang and G.S. Nolas
Department of Physics, University of South Florida
Tampa, FL 33620, USA

ABSTRACT
    Many ionic compounds with the fluorite crystal structure can accept large concentrations of substitution atoms with differing charge, producing either a large amount of vacancies or interstitials that sometimes can suppress the thermal conductivity. The intermetallic semiconductor compound Mg_2Si has the antifluorite crystal structure and has been known as a promising thermoelectric material for many years. Introducing large concentrations of vacancies or interstitials in this compound may suppress the thermal conductivity and improve the thermoelectric properties. Bearing this in mind, we synthesized Mg_2Si specimens with varying content of Mg_3Sb_2 by solid state reaction, followed by hot-press densification and characterization of their thermoelectric properties. Our results show the substitution of trivalent Sb for tetravalent Si introduces both electrons and vacancies. With the increase of Mg_3Sb_2 content both Seebeck coefficient and electrical resistivity first decrease and then increase. This can be attributed to the electron concentration increasing first and then decreasing. The lattice thermal conductivity decreases monotonically with increasing Mg_3Sb_2 content. The strain field fluctuation caused by these vacancies may play an important role in suppressing the thermal conductivity.

INTRODUCTION

    Solid-state thermoelectric energy conversion devices that generate electric power from exhaust or waste heat have attracted interest in recent years[1,2]. The efficiency of these devices depends mainly on the performance of their thermoelectric materials. The material performance is characterized by the figure of merit $Z=S^2/\kappa\rho$, where S is Seebeck coefficient, κ is thermal conductivity and ρ is electrical resistivity. For a good thermoelectric material, a large Seebeck coefficient, low electrical resistivity and low thermal conductivity are needed. However, for a given material, these three parameters are not independent. The traditional routes to optimize performance are tuning the carrier concentration to get the maximum power factor (S^2/ρ) and suppressing lattice thermal conductivity by isoelectronic alloying[3].

    Recently large improvements of Z in skutterudites were attributed to filling the voids in its structure in order to significantly suppress the lattice thermal conductivity[4,5]. This shows a new approach to improve thermoelectric materials performance: suppressing the lattice thermal conductivity by introducing interstitials. It has been found that many fluorite structure compounds, such as $Zr_{1-x}Y_xO_{2-x/2}$ ($0.18<x\leq0.30$) and $Ba_{1-x}La_xF_{2+x}$ ($0<x\leq0.5$), can accept large concentrations of substitution atoms with different charge, producing either a large concentration of interstitials or vacancies, and resulting in a glasslike thermal conductivity[6,7]. Mg_2Si is a promising thermoelectric material which has the antifluorite crystal structure. Simply stated, the difference between the antifluorite structure and the fluorite structure is the anions (cations) in the antifluorite (fluorite) structure occupy the cations (anions) sites in the fluorite (antifluorite) structure. If atoms with different valence charge were introduced in this material, large concentrations of interstitials or vacancies may be produced thus presumably suppressing the lattice thermal conductivity as compared to Mg_2Si. In this paper we report our preliminary

experimental results on introducing vacancies in the antifluorite material Mg$_2$Si by substituting Sb for Si.

EXPERIMENTAL

High-purity Mg (99.8%), Si (99.99%) and Sb (99.999%) powders were appropriately weighted to result in the chemical compositions (Mg$_2$Si)$_{1-x}$(Mg$_3$Sb$_2$)$_x$ (x=0, 0.01, 0.05 and 0.1) and mixed by mortal and pestle. To compensate for Mg evaporation loss, Mg was added in excess of 2 mol% to these stoichiometric compositions. The resulting powders were cold pressed into pellets, placed into the tungsten crucibles, sealed in silica quartz tubes under Ar atmosphere, and reacted at 850K for 24h. The resulting powders were ground, pressed into pellets and put back into the furnace for another solid state reaction at 1173K for 24h. Subsequently, the obtained compounds were ground and hot pressed at 973K for 2h under Ar flow.

The compositions of the reacted powders and sintered samples were examined by X-ray diffraction (XRD) and the lattice constants were obtained from these XRD patterns. For thermoelectric properties measurement, a 2×2×5mm^3 bar was cut from the hot pressed pellets and the Seebeck coefficient, electrical resistivity and thermal conductivity were measured simultaneously from room temperature to 10K. The measurement details were described elsewhere.[8]

RESULTS

The X-ray diffraction patterns for all (Mg$_2$Si)$_{1-x}$(Mg$_3$Sb$_2$)$_x$ (x=0, 0.01, 0.05, 0.1) specimens are shown in figure 1. Except for the small peaks (labeled with ●) attributed to MgO phase, all peaks can be indexed to the antifluorite Mg$_2$Si phase. This indicates all the specimens are composed of the Mg$_2$Si phase with a trace amount of MgO impurity. Figure 2 is the plot of the lattice parameter for (Mg$_2$Si)$_{1-x}$(Mg$_3$Sb$_2$)$_x$ as the function of Mg$_3$Sb$_2$ content calculated by Rietveld refinement. The lattice parameter increases in proportion to the Mg$_3$Sb$_2$ content, as shown in figure 2. This indicates that Sb substitutes for Si successfully. As the atom radius of Sb is bigger

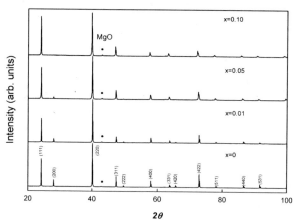

Figure 1. Powder XRD patterns for (Mg$_2$Si)$_{1-x}$(Mg$_3$Sb$_2$)$_x$ with x=0, 0.01, 0.05, 0.1.

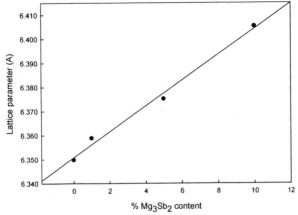

Figure 2. Lattice parameter of $(Mg_2Si)_{1-x}(Mg_3Sb_2)_x$ with content of Mg_3Sb_2.

than that of Si, by substituting Sb for Si, the lattice parameter will increase monotonically with the Mg_3Sb_2 content increasing.

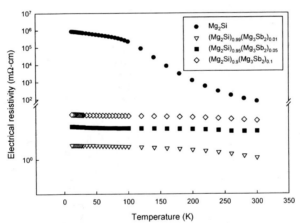

Figure 3. Temperature dependence of electrical resistivity of $(Mg_2Si)_{1-x}(Mg_3Sb_2)_x$ with x=0, 0.01, 0.05, 0.1.

The temperature dependence of electrical resistivity of three specimens along with Mg_2Si is shown in figure 3. For all the specimens, the resistivity increases with decreasing temperature, indicating a semiconductor behavior. For Mg_2Si, the electrical resistivity is quite large. However with 1atom% Mg$_3$Sb$_2$ alloying, the resistivity decreases significantly in the entire temperature range. The room temperature electrical resistivity changes from 71.6mΩ cm to 1mΩ cm between x=0 and x=0.01. Upon further increasing the content of Mg$_3$Sb$_2$, the resistivity increases.

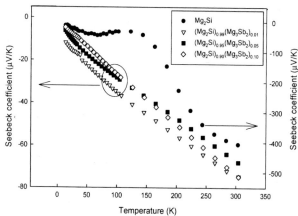

Figure 4. Seebeck coefficient VS temperature for (Mg$_2$Si)$_{1-x}$(Mg$_3$Sb$_2$)$_x$ with x=0, 0.01, 0.05, 0.1.

Figure 4 presents the Seebeck coefficient for the four specimens as function of temperature. For all the specimens, the Seebeck coefficient is negative and decreases with decreasing temperature. This indicates all specimens are n type. The absolute room temperature Seebeck coefficient of intrinsic Mg$_2$Si is rather large, 406μV/K, and decreases significantly with Mg$_3$Sb$_2$ alloying. For the specimen with 1 atom% Mg$_3$Sb$_2$ content, it is about 75μV/K. With increasing Mg$_3$Sb$_2$ content the Seebeck coefficient decreases at first, reaches the minimum when the Mg$_3$Sb$_2$ content is about 5 atom%, and increases with increasing Mg$_3$Sb$_2$ content thereafter.

Figure 5 shows the temperature dependence of thermal conductivity for the four specimens. The Mg$_3$Sb$_2$ alloying specimens show a lower thermal conductivity in the measured temperature range. In addition, with increasing Mg$_3$Sb$_2$ content the thermal conductivities decrease monotonically.

The total thermal conductivity of a solid can be written as

$$\kappa_{total} = \kappa_{lattice} + \kappa_{carrier} \tag{1}$$

According to the Wiedemann-Franz law, the electronic thermal conductivity $\kappa_{carrier}$ can be estimated by [9]

$$\kappa_{carrier} = L_0 T/\rho \tag{2}$$

where the Lorenz number $L_0 = 2.45 \times 10^{-8}$ V^2 K^{-2}. Hence, the lattice thermal conductivity can be determined by subtracting the carrier component from the measured total thermal conductivity.

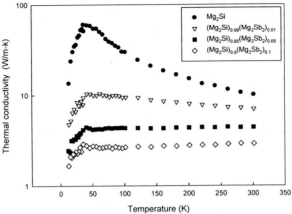

Figure 5. Temperature dependence of thermal conductivity of $(Mg_2Si)_{1-x}(Mg_3Sb_2)_x$ with x=0, 0.01, 0.05, 0.1.

Figure 6 shows the lattice thermal conductivity for the four specimens. The total thermal conductivity is mainly determined by the lattice thermal conductivity component and Mg_3Sb_2 alloying significantly suppresses this part thermal conductivity.

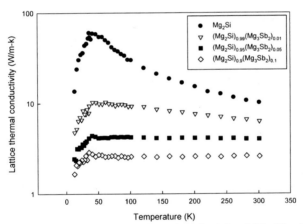

Figure 6. Temperature dependence of lattice thermal conductivity of $(Mg_2Si)_{1-x}(Mg_3Sb_2)_x$ with x=0, 0.01, 0.05, 0.1.

DISCUSSION

Mg$_2$Si is a covalent semiconducting compound. With Mg$_3$Sb$_2$ alloying, trivalent Sb substitutes for tetravalent Si in the structure. This will either introduce electrons or create Mg vacancies according to the following relation:

$$Mg_3Sb_2 \rightarrow 2Sb_{Si}^{\bullet} + V_{Mg}^{''} + 3Mg_{Mg}^{\times}$$ (3)

where, in terms of Kroger-Vink notation $Sb_{Si}^{\bullet}$ represents a Sb^{3-} ion that occupies a Si^{4+} ion site (single positive charge), $V_{Mg}^{''}$ is a doubly charged (negative) Mg vacancy, and $Mg_{Mg}^{\times}$ is an Mg^{2+} ion on an Mg site (neutral charge)[10].

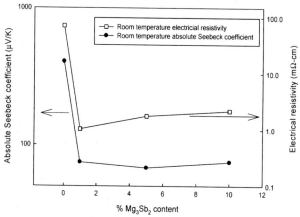

Figure 7. Room temperature electrical resistivity and absolute Seebeck coefficient versus Mg$_3$Sb$_2$ content.

If all Sb^{3-} substituting for Si^{4+} only generated vacancies on the Mg site, the carrier concentration for all the specimens should be almost the same (intrinsic carrier concentration). Neither Seebeck coefficient nor electrical conductivity should change significantly. This contradicts our experimental results as both Seebeck coefficient and resistivity change greatly with Mg$_3$Sb$_2$ alloying. For n type semiconductors, the Seebeck coefficient and electrical resistivity can be expressed as

$$S = -[(E_C - E_F)/eT + 2k_B/e]$$ (4)

$$\rho = 1/(ne\mu)$$ (5)

where k$_B$ is the Boltzmann constant, E$_C$ is the minimum of the conduction band, E$_F$ is the Fermi level, n is the carrier concentration, μ is the carrier mobility, and T is the absolute temperature[9].

Increasing the electron concentration will result in the Fermi level normally shifting toward the conduction band, which leads to a smaller difference of $E_C - E_F$ and a smaller absolute Seebeck coefficient. Increasing the electron concentration also decreases the electrical resistivity if the mobility does not degrade extensively. Figure 7 shows the room temperature electrical resistivity and absolute Seebeck coefficient of all the measured samples as function of Mg_3Sb_2 content. Consider the dependence of Seebeck coefficient on Mg_3Sb_2 content. The Seebeck coefficient first decreases with Mg_3Sb_2 content from 0 to 5 atom%, indicating an increase in carrier concentration. Upon increasing Mg_3Sb_2 content further from 5 to 10 atom%, the Seebeck coefficient then increases, indicating a decrease in carrier concentration. The resistivity change agrees with this conclusion very well for the specimens with 1 atom% and 10 atom% Mg_3Sb_2 alloying. However, for the specimens with a Mg_3Sb_2 content of 5 atom%, the resistivity increases while the Seebeck coefficient decreases. This can be explained by considering the degradation of mobility with increasing Mg_3Sb_2 content. In fact, defects will enhance scattering of carrier and result in a degradation of the mobility. For the specimens with a Mg_3Sb_2 content varying from 1 atom% to 5 atom%, the Seebeck coefficient changes slightly, from -75 μV/K to - 68 μV/K. This indicates the electron concentration increases slightly. If the decrease of resistivity because of carrier concentration increase is smaller than the increase of resistivity because of degradation of mobility, the Seebeck coefficient will decrease while the resistivity will increase.

If the substitution Sb^{3-} for Si^{4+} only introduces electrons, in a rigid band type model, the carrier concentration can be estimated by assuming one Sb^{3-} substitution for Si^{4+} generating one electron. For the specimens with 1 atom%, 5 atom% and 10 atom% Mg_3Sb_2 alloying, the calculated carrier concentrations were $6.2 \times 10^{20} cm^{-3}$, 3.0×10^{21} cm^{-3} and $5.7 \times 10^{21} cm^{-3}$, respectively. However, for the specimen with the 5 tom% Mg_3Sb_2 content, the room temperature Seebeck coefficient is -68 μV/K and the temperature dependence of electrical resistivity shows semiconducting behavior. Therefore, it is reasonable to estimate that the carrier concentration is approximately $10^{20} cm^{-3}$, one order of magnitude lower than that estimated[11]. This indicates only a small amount of Sb substitution contributes to the increase in carrier concentration, while most generate vacancies on the Mg sites. If we ignore the intrinsic vacancies and assume 90% Sb substitution for Si generating vacancies according to Eq. (3), for the specimen with 10 atom% Mg_3Sb_2 substitution, the Mg vacancy concentration ($[V''_{Mg}]$) amounts to 4%. In $Zr_{1-x}Y_xO_{2-x/2}$ (x=0.2), large amount of vacancies are generated by substituting Y for Zr and the lattice thermal conductivity is greatly suppressed[12]. Y. Klemens et al[13] attributed this mainly to scattering by the oxygen vacancies and emphasized that the oxygen vacancy is a much stronger phonon scatterer than the Y^{3+} cations because the mass and size difference between a solid solution of Y^{3+} cations and Zr^{4+} cations is relatively small. In Mg_2Si with Mg_3Sb_2 alloying, except for the mass and strain field fluctuation caused by Sb substitution for Si contributing to the suppress the lattice thermal conductivity, the large amount of vacancies generated must also play an important role in suppressing the lattice thermal conductivity. As Mg is relatively light, the mass fluctuation might not be as large a factor. Strain field fluctuation caused by these vacancies may play a more important role in scattering phonon. To clarify this, further quantitatively calculation is needed and this, along with Hall measurement, is now underway in our laboratory.

CONCLUSION
    Single phase M_2Si with varying concentrations of Mg_3Sb_2 were prepared and their low temperature transport properties were measured. All samples show semiconducting behavior. The Seebeck coefficient and electrical resistivity depends on the Mg_3Sb_2 content indicating that

the carrier concentration increases with increasing Mg_3Sb_2 content from 0 to 5atom% and then decreases with further increasing Mg_3Sb_2 content. The lattice thermal conductivity is suppressed greatly with Mg_3Sb_2 alloying by introducing phonon scattering vacancies at the Mg sites as well as solid solution of Sb at Si sites. The strain field fluctuation caused by these vacancies may play an important role in completely understanding the transport properties. Introducing a large amount of vacancies may indicate a way to improve the thermoelectric properties of the antifluorite structured material.

The authors acknowledge support by GM and DOE under corporate agreement DE-FC26–04NT42278.

REFERENCES

[1]F.J. di Salvo, Science 285 (5428), 703, (1999).
[2]X. Ma and S.B. Riffat, Appl Therm Eng 23 (8), 913, (2003).
[3]D.M. Rowe, Editor, CRC Handbook of thermoelectrics, CRC Press, Boca Raton, FL (1995).
[4]G.S. Nolas, D.T. Morelli and T.M. Tritt, Annu. Rev. Mater. Sci. 29, 89, (1999).
[5]L. D. Chen, T. Kawahara, X. F. Tang, T. Goto, and T. Hirai, J. Appl. Phys. 90, 1864 (1996).
[6]David G. Cahill, S. K. Watson and R. O. Pohl, Phys. Rev. B, 46 (10), 6131, (1992).
[7]F. J. Walker and A. C. Anderson, Phys. Rev. B, 29 (10), 5881, (1984).
[8]J. Martin, S. Erickson, G. S. Nolas P. Alboni and T.M. Tritt, J. Appl. Phy., 99, 044903 (2006)
[9]G. S. Nolas, J. Sharp, and H. J. Goldsmid, Thermoelectrics: Basic principles and new materials developments, Springer New York (2000)
[10]S. M. Allen and E. L. Thomas, The Structure of Materials, John Wiley & Sons, New York, 1999.
[11]J. Tani and H. Kido, Phy. B, 364, 218 (2005)
[12]J. Bisson, D. Fournier, M. Poulain, O. Lavigne, and R. Mévrel, J. Am. Ceram. Soc. 83 (8), 1993, (2000)
[13]P. G. Klemens, "Thermal conductivity of Zirconia"; pp. 209-20 in Thermal Conductivity, Vol. 23. Technomic, Lancaster, PA (1996)

THERMOELECTRIC PERFORMANCE OF DOPED SrO(SrTiO₃)ₙ (n = 1, 2) RUDDLESDEN-POPPER PHASES

Kyu Hyoung Lee
CREST, Japan Science and Technology Agency
4-1-8, Honmachi, Kawaguchi, 332-0012, Japan

Yi Feng Wang
Graduate School of Engineering, Nagoya University
Nagoya, 464-8603, Japan

Hiromichi Ohta, Kunihito Koumoto
CREST, Japan Science and Technology Agency
4-1-8, Honmachi, Kawaguchi, 332-0012, Japan
Graduate School of Engineering, Nagoya University
Nagoya, 464-8603, Japan

ABSTRACT
        We have investigated the thermoelectric properties and crystallographic features of Ruddlesden-Popper phase Ca- and/or Nb-doped SrO(SrTiO₃)ₙ (n = 1, 2) to elucidate their potential as thermoelectric materials and clarify the influence of crystal structure on the carrier effective mass in Ti-based oxides containing TiO_6 octahedra. Significant reduction in the lattice thermal conductivity by suppression of the mean free path of phonons possibly associated with the presence of internal sublattice SrO/(SrTiO₃)ₙ interfaces was observed. It was also found that large Seebeck coefficients of Ti-based oxides can be attributed to the high value of carrier effective mass, which originates from the high symmetry of the TiO_6 octahedra. The overall ZT values obtained in the present study were in the range of 0.09 – 0.15 at 1000 K.

INTRODUCTION
        Ti-based oxides including perovskite-type $ATiO_3$ (A = Ca, Sr, and Ba) would be the most promising n-type semiconductors for high temperature thermoelectric (TE) applications owing to rather high TE performance as well as good chemical and thermal resistances at high temperatures.[1-5] In our previous study,[1] the cubic perovskite-type Nb-doped SrTiO₃ has been found to exhibit the largest dimensionless TE figure of merit, $ZT_{1000 K} = 0.37$ ($ZT = S^2\sigma T/\kappa$, where Z, T, S, σ, and κ are the figure of merit, absolute temperature, Seebeck coefficient, electrical conductivity, and thermal conductivity, respectively) among n-type TE oxides ever reported. This large ZT value can be attributed to its large S value, originated from the large carrier effective mass (m^*) even at high carrier concentration (n_c). On the other hand, much smaller S and m^* values are found in other Ti-based oxides with deformed TiO_6 octahedra such as Nb-doped anatase TiO_2.[6] Because the S and m^* reflect the density of states (DOS) of the bottom of the conduction band (CB), this large difference in S and m^* values between cubic perovskite-type SrTiO₃ and anatase TiO_2 is considered to be related with the feature of DOS of the bottom of each CB. The bottom of the CB of the SrTiO₃ is, as well known, composed of Ti 3d-t_{2g} triple degenerated states,[7] however, Ti 3d-t_{2g} states in anatase-TiO_2 split into two different states (dispersion-less state, d_{xy} and the other two energy-elevated dispersive states, d_{yz} and d_{zx})[8] due to crystal field splitting in the presence of deformed TiO_6 octahedra and this characteristic

feature of Ti 3d-t$_{2g}$ orbital of anatase TiO$_2$ might be responsible for small S (small m*) values. Attention has been paid to the clarification of the effects of crystal structure on the transport parameters, including S and m* for the realization of high TE performance in Ti-based oxides.

Layered perovskite-type SrO(SrTiO$_3$)$_n$ (or Sr$_{n+1}$Ti$_n$O$_{3n+1}$, n = integer) Ruddlesden-Popper (RP) phases are very good candidates for investigation of the relationships between crystal structure and transport parameters, having the (SrTiO$_3$)$_n$ layers with deformed TiO$_6$, in Ti-based oxides.[9,10] These oxides also have the possibilities of exhibiting low κ, very attractive property for TE applications, owing to their superlattice period of SrO/(SrTiO$_3$)$_n$ and maintaining the favorable electronic properties of SrTiO$_3$. In the present study, we investigated the TE properties of Nb-doped Sr$_{n+1}$Ti$_n$O$_{3n+1}$ (n = 1, 2) and Ca- and Nb-doped n = 2 Sr$_3$Ti$_2$O$_7$, and demonstrated the origin of TE properties in Ti-based metal oxides containing TiO$_6$ octahedra from the viewpoint of the crystal structure.

EXPERIMENTAL

0 – 20 at.% Nb-doped Sr$_{n+1}$Ti$_n$O$_{3n+1}$ (n = 1, 2) and 10 at.% Ca- and 5 – 20 at.% Nb-doped n = 2 Sr$_3$Ti$_2$O$_7$ powders were prepared by the solid-state reaction using high purity (> 99.9 %) SrCO$_3$, CaCO$_3$, TiO$_2$ and Nb$_2$O$_5$ powders. The powders were mixed for 1 h in a planetary ball mill and calcined for 12 h at 1200 °C in air for decarbonization and homogenization of the final samples. In order to form the RP phases and generate the electron carriers through the reduction of Ti^{4+} to Ti^{3+} by doping of Nb^{5+}, the powder was heated at 1400 – 1475 °C for 1 – 2 h in a carbon crucible under an Ar atmosphere. Then, highly dense polycrystalline ceramic samples were fabricated by conventional hot pressing (36 MPa and 1400 – 1475 °C for 1 h in an Ar flow). Structural parameters obtained from the X-ray diffraction data were refined by the Rietveld method using the RIETAN-2000 program.[11] The σ and S values were measured from 300 to 1000 K by a conventional dc 4-probe method and a steady state method, respectively, in flowing Ar. The n$_c$ values were measured with the van der Pauw configuration under vacuum. The κ values were calculated by separate measurements with differential scanning calorimetry for heat capacity and a laser-flash method for thermal diffusivity under vacuum and bulk density (~ 97 % relative density).

RESULT AND DISCUSSION

The powder X-ray diffraction patterns for 0 – 20 at.% Nb-doped Sr$_{n+1}$Ti$_n$O$_{3n+1}$ (n = 1, 2) i.e. Sr$_2$Ti$_{1-x}$Nb$_x$O$_4$ and Sr$_3$(Ti$_{1-x}$Nb$_x$)$_2$O$_7$ (x = 0 – 20), and 10 at.% Ca- and 5 – 20 at.% Nb-doped n = 2 Sr$_3$Ti$_2$O$_7$ i.e. (Sr$_{0.9}$Ca$_{0.1}$)$_3$(Ti$_{1-x}$Nb$_x$)$_2$O$_7$ (x = 0.05 – 20) compositions are shown in Fig. 1. For Nb-doped samples (Fig. 1(a)), peaks for other phases were not detected in n = 2 compositions, while a small amount of n = 2 phase was found as a second phase in all n = 1 samples, suggesting that n = 2 is the most stable among the Sr-Ti-O system RP phases[12] and the solubility limit of Nb^{5+} ion on Ti-site is about 20 at.%. On the other hand, Sr$_4$Nb$_2$O$_9$ was produced as the secondary phase in 10 at.% Ca- and 20 at.% Nb-doped SrO(SrTiO$_3$)$_2$ compositions (Fig. 1(b)). This result could be explained in terms of a tolerance factor, which is well known as the parameter of the perovskite-type structure stability. From the fact that the tolerance factor of SrTiO$_3$ is about 1.0 and it is decreased by doping large Nb^{5+} ion (r = 64 pm, coordination number = 6) on Ti^{4+}- (r = 60.5 pm, coordination number = 6) site as well as small Ca^{2+} ion (r = 134 pm, coordination number = 12) on Sr^{2+}- (r = 144 pm, coordination number = 12) site,[13] it can be understood that solubility limit of Nb^{5+} ion on Ti-site can be decreased in the Ca-doped compositions.

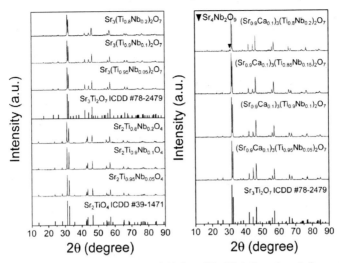

Figure 1. Powder X-ray diffraction patterns of (a) Sr$_{n+1}$(Ti$_{1-x}$Nb$_x$)$_n$O$_{3n+1}$ (n = 1, 2, x = 0.05 – 0.2) and (b) (Sr$_{0.9}$Ca$_{0.1}$)$_3$(Ti$_{1-x}$Nb$_x$)$_2$O$_7$ (x = 0.05 – 0.2)

Rietveld refinement was conducted based on the space group I4/mmm (No. 139). The reliability factor R$_{wp}$ was ~ 10 % in all compounds, and the obtained crystallographic data are given in Fig. 2.

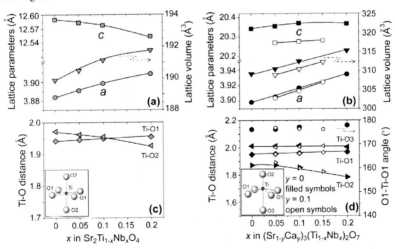

Figure 2. Variation of crystallographic parameters for (a) Sr$_2$Ti$_{1-x}$Nb$_x$O$_4$ (x = 0 – 0.2) and (b) (Sr$_{1-y}$Ca$_y$)$_3$(Ti$_{1-x}$Nb$_x$)$_2$O$_7$ (x = 0 – 0.2, y = 0, 0.1)

As shown in Fig. 2(a) and (c), the lattice volume (V) increases gradually with an increase in Nb content and decreases in Ca-doped compositions. All of findings can be attributed to the fact that the radius of the Ca^{2+} ion is smaller than that of Sr^{2+} ion and Nb^{5+} ion is slightly larger than that of Ti^{4+} ion, and larger Ti^{3+} ions (r = 67 pm, coordination number = 6) is created by Nb doping. The structure of SrO(SrTiO₃)ₙ can be regarded as the alternative stacking of perovskite block (SrTiO₃)ₙ layers and SrO rock salt type layers, however, the SrTiO₃ in RP phases is not cubic symmetry, which implies the presence of deformed TiO₆ octahedra owing to the stress from the SrO layers. The insets of Fig. 2(b) and (d) show the schematic structure of TiO₆ octahedra in perovskite layers of RP phases. As shown in Fig. 2(b), there are two different Ti-O bonds in n = 1 Sr₂TiO₄: the shorter Ti-O1 in the ab plane and the longer Ti-O2 along the c-axis. While n = 2 Sr₃Ti₂O₇ has three different Ti-O bonds, and the Ti ion is situated slightly above the O1 ions along the c-axis. As a result, the Ti-O1 layers are not flat, but a little corrugated. In order to explore the shape of TiO₆ octahedra, the length of Ti-O bonds of n = 1 and 2 compounds and O1-Ti-O1 angle of n = 2 compounds were calculated and the obtained data were shown in Fig. 2(b) and (d). The feature of TiO₆ octahedra could be changed by doping, however, TiO₆ octahedra were still remained deformed shape in all compositions.

Figure 3(a) shows the temperature dependence of the σ for n = 1 Sr₂Ti₁₋ₓNbₓO₄ (x = 0.05 – 0.2) and n = 2 (Sr₁₋ᵧCaᵧ)₃(Ti₁₋ₓNbₓ)₂O₇ (x = 0.05 – 0.2, y = 0, 0.1). The σ of the cubic perovskite-type 20 at.% Nb-doped SrTiO₃ polycrystalline samples are shown for comparison.[14]

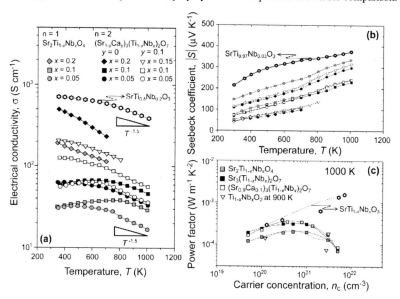

Figure 3. Temperature dependence of (a) electrical conductivity and (b) Seebeck coefficient, and (c) estimated power factor with carrier concentration for Sr₂Ti₁₋ₓNbₓO₄ (x = 0.05 – 0.2) and (Sr₁₋ᵧCaᵧ)₃(Ti₁₋ₓNbₓ)₂O₇ (x = 0.05 – 0.2, y = 0, 0.1). Data for Nb-doped SrTiO₃ and TiO₂ are taken from refs. 1, 2, 14, and 6.

Since the doped Nb atom acts as an electron donor, the n_c increases with increasing doping amount, leading to the increase in σ. The σ values depend only slightly on temperature at low temperatures, whereas, reaching higher temperatures ($\sim$ 750 K), a strong temperature dependence in the form of a power law, $\sigma \propto T^{-1.5}$, is observed. From the results of similar power law dependences in perovskite-type SrTiO₃ systems,[1,3,15] which have indicated that the acoustic phonon scattering mechanism is predominant, it has been considered that the electrical conduction in RP phases should take place dominantly within the perovskite layers. The lower σ of RP phases than the perovskite-type ones is considered to be mainly due to the insulating SrO layers randomly distributed in polycrystalline samples. Figure 3(b) and (c) show the temperature dependence of the absolute values of S and the n_c dependence of the estimated power factor (PF). The S of the cubic perovskite-type 3 at.% Nb-doped SrTiO₃ polycrystalline samples are shown for comparison. All samples have negative S, with a gradual increase in magnitude with temperature, indicating that the samples are n-type degenerate semiconductors. In spite of the lower n_c values of 5 at.% Nb-doped RP phases ($n_c \sim 3 \times 10^{20}$ cm^{-3}) compared to those of SrTiO₃ phases ($n_c \sim 5 \times 10^{20}$ cm^{-3}), the S values of the former are lower than those for the latter. In order to clarify this difference in S values, we have calculated the m^* values, which is one of the main factors determining S, and examined the transport and structural parameters affecting m^* in Ti-based oxides. Details of the calculation of m^* are described elsewhere.[16]

Figure 4 shows the m^* values as a function of Ti-Ti distance and n_c at 1000 K. Values for the cubic perovskite-type La- and Nb-doped SrTiO₃ single crystals[2] and epitaxial films[1] (at 1000 K) and Nb-doped anatase TiO₂ epitaxial films[6] (at 900 K) are shown for comparison. In many TE materials, the variation behavior of m^* can be explained according to the Kane model,[17] which represents an increase in m^* with increasing charge carrier concentration. However, the m^* values for Ti-based metal oxides appear to be independent of n_c, as shown in Fig. 4 (b). In our previous studies on cubic perovskite-type La- and Nb-doped SrTiO₃ systems,[1] we proposed the Ti-Ti distance as the parameter responsible for determining m^*, and it was found that the increase in m^* by Nb doping did not originate from an increase in n_c but from an increase in the distance between two neighboring Ti ions, which leads to a decrease in the overlap between Ti 3d-t₂g orbitals. Increases in S by lattice expansion (enlargement of Ti-Ti distance) are observed in other cubic perovskite-type Ba-substituted and Y-doped SrTiO₃ systems,[3,18] however, this phenomenon is not effective in RP phases with distorted TiO₆ octahedra. Thus, the PF values of RP phases and anatase TiO₂ were much lower than that of cubic perovskite-type Nb-doped SrTiO₃, especially at high n_c, as shown in Fig. 3(c). The resulting m^* values of RP phases are 1.8 – 2.4 m_0 and show great deviation from the linear relation of cubic perovskite-type SrTiO₃ (Fig. 4(a)). Very recently, we have investigated the TE properties of heavily Nb-doped anatase TiO₂ epitaxial films, which also contain distorted TiO₆ octahedra.[6] The carrier generation mechanism and transport properties were basically similar to those of Nb-doped SrTiO₃, whereas m^* was an order of magnitude smaller than that of Nb-doped SrTiO₃. As above mentioned, small m^* values for RP phases and TiO₂ are due to the crystal field splitting of Ti 3d-t₂g orbitals (splitting of degeneracy, Fig. 4(c))[8,19] by the presence of deformed TiO₆ octahedra and could be responsible for low S. Thus, in order to achieve a high TE performance in Ti-based metal oxides, it is highly desirable that both high symmetry for the TiO₆ octahedra and large length for the Ti-Ti bonds, as in SrTiO₃ perovskites, should be maintained. The present findings may provide highly useful information for the realization of high TE performance in Ti-based metal oxides.

On the other hand, a slight increase in S was observed in Ca-doped compounds (Fig. 3(b)). From the results that the TiO₆ octahedra become more symmetric by Ca doping (Fig. 2(d), the increase in S by Ca doping is considered to be caused by the symmetry of the TiO_6 octahedra.

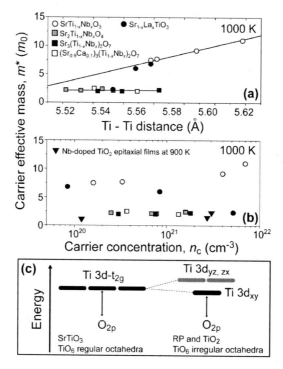

Figure 4. Carrier effective mass as functions of (a) Ti-Ti distance along [110] and (b) carrier concentration at 1000 K, and (c)Schematic energy diagram of Ti 3d-t_{2g} orbitals. Data for La- and Nb-doped $SrTiO_3$ (at 1000 K) and Nb-doped anatase TiO_2 epitaxial films (at 900 K) are taken from refs. 1, 2, and 6.

Figure 5 shows the temperature dependence of κ for 5 at.% Nb-doped n = 1 Sr_2TiO_4 and n = 2 $Sr_3Ti_2O_7$ and $(Sr_{0.9}Ca_{0.1})_3Ti_2O_7$. The κ of the cubic perovskite-type 20 at.% Nb-doped $SrTiO_3$ polycrystalline samples are shown for comparison.[14] The total thermal conductivity ($κ_{tot}$) can be expressed by the sum of lattice ($κ_{lat}$) and the electronic contribution ($κ_{ele}$). In our case, $κ_{ele}$ increases with increasing Nb content and by Ca doping, owing to the increase in n_c, however, the $κ_{ele}$ values estimated by the Wiedemann-Franz law are very small ($κ_{ele}$ ~ 0.3 W m⁻¹ K⁻¹) compared to $κ_{tot}$, which indicates that the phonon contribution is predominant. In comparison with the cubic perovskite-type Nb-doped $SrTiO_3$ polycrystalline ceramics, $κ_{lat}$ decreased in value by more than ~ 50 % (~ 5 W m⁻¹ K⁻¹) at room temperature and ~ 20 % (~ 2.7 W m⁻¹ K⁻¹) at 1000 K, respectively. This reduction in the κ value would have been caused by the enhancement of phonon scattering at the $SrO/(SrTiO_3)_n$ internal interfaces and additional reduction by Ca doping was found due to mass-defect phonon scattering between Ca (M_{Ca} = 40) and Sr (M_{Sr} = 88) at low temperatures. The κ values for Ca-doped compositions show weak temperature dependence, and κ values were higher than those of undoped ones above 700 K.

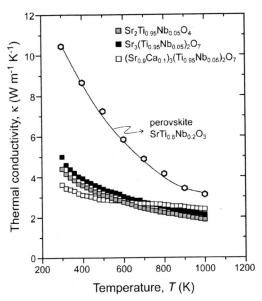

Figure 5. Temperature dependence of thermal conductivity for 5 at.% Nb-doped n = 1 Sr$_2$TiO$_4$ and n = 2 Sr$_3$Ti$_2$O$_7$ and (Sr$_{0.9}$Ca$_{0.1}$)$_3$Ti$_2$O$_7$. The thermal conductivity data for cubic perovskite-type 20 at.% Nb-doped SrTiO$_3$ polycrystalline ceramics are taken from the data in ref. 14.

From the σ, S, and κ values, we have calculated the ZT values for Sr$_2$Ti$_{1-x}$Nb$_x$O$_4$ (x = 0.05 – 0.2) and (Sr$_{1-y}$Ca$_y$)$_3$(Ti$_{1-x}$Nb$_x$)$_2$O$_7$ (x = 0.05 – 0.2, y = 0, 0.1) and present them in Fig. 6. The ZT values of the cubic perovskite-type 20 at.% Nb-doped SrTiO$_3$ polycrystalline samples are shown for comparison.[14] Overall ZT values of RP phases were in the range of 0.09 – 0.15 at 1000 K, and the largest ZT value (0.15 at 1000 K) was observed for 10 at. % Ca- and 5 at.% Nb-doped n = 2 Sr$_3$Ti$_2$O$_7$. In spite of their low κ, ZT values of RP phases were lower than that of cubic perovskite-type Nb-doped SrTiO$_3$ (0.37 at 1000 K). As mentioned above, the key reasons for relatively low ZT of RP phases are firstly the low σ caused by the insertion of insulating SrO layers into perovskite structures and secondly the small S (low m*), possibly caused by splitting of degeneracy due to the deformation of TiO$_6$ octahedra in perovskite layers. The following points remain to be further clarified in future investigations in order to obtain the maximum TE efficiency in RP phases: the optimum composition to realize high symmetry TiO$_6$ octahedra and large length for Ti-Ti bonds in perovskite layers, and processes for producing textured (crystal-axis-oriented) ceramics and epitaxial thin films.

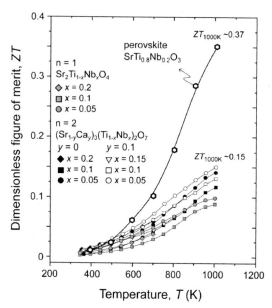

Figure 6. Temperature dependence of dimensionless thermoelectric figures of merit for $Sr_2Ti_{1-x}Nb_xO_4$ (x = 0.05 – 0.2) and $(Sr_{1-y}Ca_y)_3(Ti_{1-x}Nb_x)_2O_7$ (x = 0.05 – 0.2, y = 0, 0.1). Data for cubic perovskite-type 20 at.% Nb-doped SrTiO₃ polycrystalline ceramics are taken from the data in ref. 14.

CONCLUSION

We investigated the TE properties and relationships between the transport parameters and the crystallographic parameters for layered perovskite-type $Sr_2Ti_{1-x}Nb_xO_4$ (x = 0.05 – 0.2) and $(Sr_{1-x}Ca_y)_3(Ti_{1-x}Nb_x)_2O_7$ (x = 0.05 – 0.2, y = 0, 0.1) polycrystalline RP phases. It was found that large S values could be obtained by enhancement of the symmetry of the TiO_6 octahedra resulting in increases in m*, suggesting the both high TiO_6 symmetry and adequate Ti-Ti distance are required for high TE performance Ti-based metal oxides.

In addition to the κ reduction by phonon scattering at SrO/(SrTiO₃)ₙ interfaces of the inherent superlattice structure, there was additional reduction achieved due to the mass-defect phonon scattering between Ca (M_{Ca} = 40) and Sr (M_{Sr} = 88) at low temperatures. The largest ZT value observed (0.15 at 1000 K) was obtained for 10 at. % Ca- and 5 at.% Nb-doped n = 2 $Sr_3Ti_2O_7$.

REFERENCES

[1]S. Ohta, T. Nomura, H. Ohta, M. Hirano, H. Hosono, and K. Koumoto, "Large Thermoelectric Performance of Heavily Nb-Doped SrTiO₃ Epitaxial Film at High Temperature," *Appl. Phys. Lett.*, **87**, 092108 (2005).

[2] S. Ohta, T. Nomura, H. Ohta, and K. Koumoto, "High-Temperature Carrier Transport and Thermoelectric Properties of Heavily La- or Nb-Doped SrTiO₃ Single Crystals," *J. Appl. Phys.*, **97**, 034106 (2005).

[3] H. Muta, K. Kurosaki, and S. Yamanaka, "Thermoelectric Properties of Doped BaTiO₃-SrTiO₃ Solid Solution," *J. Alloy. Comp.*, **368**, 22-24 (2004).

[4] H. Muta, A. Ieda, K. Kurosaki, and S. Yamanaka, "Substitution Effect on The Thermoelectric Properties of Alkaline Earth Titanate," *Mater. Lett.*, **58**, 3868-3871 (2004).

[5] H. P. R. Frederikse, W. R. Thurber, and W. R. Hosler, "Electronic Transport in Strontium Titanate," *Phys. Rev.*, **134**, A442-A445 (1964).

[6] D. Kurita, S. Ohta, K. Sugiura, H. Ohta, and K. Koumoto, "Carrier Generation and Transport Properties of Heavily Nb-Doped Anatase TiO₂ Epitaxial Films at High Temperatures," *J. Appl. Phys.*, **100**, 096105 (2006).

[7] K. Uchida, and S. Tsuneyuki, "First-Principles Calculations of Carrier-Doping Effects in SrTiO₃," *Phys. Rev. B*, **68**, 174107 (2003).

[8] R. Asahi, Y. Taga, W. Mannstadt, and A. J. Freeman, "Electronic and Optical Properties of Anatase TiO₂," *Phys. Rev. B*, **61**, 7459-7465 (2000).

[9] S. N. Ruddlesden, and P. Popper, "New Compounds of The K₂NiF₄ Type," *Acta Crystallogr.*, **10**, 538-539 (1957).

[10] S. N. Ruddlesden, and P. Popper, "The Compound Sr₃Ti₂O₇ and Its Structure," *Acta Crystallogr.*, **11**, 54-55 (1958).

[11] F. Izumi, and T. Ikeda, "A Rietveld-Analysis Program RIETAN-98 and Its Applications to Zeolites," *Mater. Sci. Forum*, **321**, 198-203 (2000).

[12] A. McCoy, R. W. Grimes, and W. E. Lee, "Phase Stability and Interfacial Structures in The SrO-SrTiO₃ System," *Philo. Mag.*, **75**, 833-846 (1997).

[13] R. D. Shannon, "Revised Effective Ionic Radii and Systematic Studies of Interatomic Distances in Halides and Chalcogenides," *Acta Crystallogr. Sec. A*, **32**, 751-767 (1976).

[14] S. Ohta, H. Ohta, and K. Koumoto, "Grain Size Dependence of Thermoelectric Performance of Nb-Doped SrTiO₃ Polycrystals," *J. Ceram. Soc. Jpn.*, **114**, 102-105 (2006).

[15] R. Moos, and K. H. Härdtl, "Electronic Transport Properties of Sr₁₋ₓLaₓTiO₃ Ceramics," *J. Appl. Phys.*, **80**, 393-400 (1996).

[16] V. I. Fistul, *Heavily Doped Semiconductors* (Plenum Press, 1969).

[17] E. O. Kane, "Band Structure of Indium Antimonide," *J. Phys. Chem. Solids*, **1**, 249-261 (1957).

[18] S. Hui, and A. Petric, "Electrical Properties of Yttrium-Doped Strontium Titanate under Reducing Conditions," *J. Electrochem. Soc.*, **149**, J1-J10 (2002).

[19] T. Schimizu, and T. Yamaguchi, "Band Offsets Design with Quantum-Well Gate Insulating Structures," *Appl. Phys. Lett.*, **85**, 1167-1168 (2004).

GROWTH OF $Bi_2Ca_2Co_{1.69}O_x$ COBALTITE RODS BY LASER FLOATING ZONE METHOD

E. Guilmeau,[1] A Sotelo,[2] M. A. Madre,[2] D. Chateigner,[1] and D. Grebille[1]

[1]CRISMAT Laboratory, UMR 6508 CNRS-ENSICAEN. 6 Bd Maréchal Juin. 14050 CAEN cedex, E-mail: emmanuel.guilmeau@ensicaen.fr

[2]Instituto de Ciencia de Materiales de Aragón (UZ-CSIC). Dpto. Ciencia y Tecnología de Materiales y Fluidos. C/Mª de Luna, 3. 50018 Zaragoza

ABSTRACT

This paper reports the synthesis and characterization of Bi-based misfit cobaltite rods. The specimens have been processed through the Laser Floating Zone method in air. The electrical resistivity (ρ) of directionally solidified rods were determined and correlated with the textural and microstructural features (neutron diffraction and scanning electron microscopy). This work is the first step of our researches in the field of Bi-cobaltite as-synthesized by the Laser Floating Zone method, and gives an interesting approach with the view to engineering their properties and synthesizing other promising thermoelectric compounds.

1. INTRODUCTION

Since the discovery of large thermoelectricity in Na_xCoO_2,[1] enthusiastic efforts have been devoted to explore new Co oxides exhibiting high thermoelectric performances, and some layered cobaltites, such as $[Ca_2CoO_3][CoO_2]_{1.62}$ and $[Bi_{0.87}SrO_2]_2[CoO_2]_{1.82}$ were found to exhibit good thermoelectric (TE) properties as well.[2-4]

The crystal structure of these layered cobaltites is composed of an alternate stacking of a common conductive CdI_2-type CoO_2 layer with a two-dimensional triangular lattice and a block layer, composed of one (in Na_xCoO_2) to several insulating rock-salt-type (RS) layer (in $[Bi_{0.87}SrO_2]_2[CoO_2]_{1.82}$ or $[Bi_{0.81}CaO_2]_2[CoO_2]_{1.69}$)[5-7]. Derived from Na_xCoO_2, the structure of these cobalt oxides consist of single hexagonal CoO_2 layers stacked with quadruple rock-salt layers composed of double [Bi-O] and [Ca-O] layers. The two RS and CoO_2 layers have common a and c axes, while the b-axis lengths of the two layers are different. Due to their high structural anisotropy, the alignment of plate-like grains by mechanical and/or chemical processes is necessary to attain macroscopic properties comparable to the intrinsic crystallographic ones. The preferential grain orientation is expected to improve the transport properties of the bulk material and to reach, if possible, the TE properties of the single crystal. Here is reported the processing of long textured Bi-based cobaltite bulk ceramics by the laser floating zone (LFZ) technique[8-9]. Long lengths (more than 25 cm) of textured materials have been processed. The microstructural characterization of the textured ceramics has been performed by scanning electron microscopy (SEM), and the texture was determined by neutron diffraction.

2. EXPERIMENTAL

Polycrystalline ceramics with the initial composition $Bi_2Ca_2Co_{1.7}O_x$, $Bi_2Sr_2Co_{1.8}O_x$, and $Bi_2(Ca,Sr)_2Co_{1.75}O_x$ were prepared by the conventional solid-state synthesis technique from commercial Bi_2O_3 (Panreac, 98 + %), $SrCO_3$ (Panreac, 98 + %), $CaCO_3$ (Panreac, 98.5 + %) and Co_2O_3 (Aldrich, 98 + %) powders.

These powders were weighed in the adequate proportions, mixed, milled in an agate ball mill for 30 minutes at 60 rpm, and calcined at 750°C for 12 h in air to assure the carbonates decomposition. The resulting powder was then carefully ground in an agate mortar, followed by a ball milling for 30 minutes at 60 rpm to assure good homogeneity of the mixture. The obtained product was then introduced in a furnace at 800°C for 12 h in air, reground and ball milled to obtain a fine powder, which was isostatically pressed at 200 MPa in order to obtain green ceramic cylinders. These cylinders have been then heated at 800°C for several hours, and quenched at room temperature in air, to improve their mechanical properties as they have to be used as feed in a LFZ device[8] equipped with a power Nd:YAG continuous laser (1.06 µm). The growth was performed downwards at growth rates of 15, 30 and 50 mm/h for each composition, with a relative rotation, between feed and seed, of 15 rpm, leading to long (more than 25 cm) and textured cylindrical rods.

The texture of the hot-forged sample was determined from neutron diffraction spectra. For that purpose, the cylindrical rod was cut in several pieces of 10 mm in length. They were paste ones beside the others to form an approximate rod of 10 mm in diameter and 10 mm in length. With this almost symmetric configuration, the whole volume of the sample is analyzed by the neutron beam. A curved position-sensitive detector coupled to a tilt angle (χ) scan allowed the whole diffraction pattern treatment in the combined Rietveld-WIMV algorithm, implemented in the MAUD software.[10-11] X-ray structural determination performed on the single crystal[12] was used to describe the 3D structural model of the $[Bi_{0.81}CaO_2]_2[CoO_2]_{1.69}$ phase in the MAUD software using the supercell description, and to determine through an iterative methodology the texture of the cobaltite. Experiments were carried out on the D1B neutron line at the Institut Laüe Langevin, Grenoble. The neutron wavelength is monochromatised to $\lambda=2.523$ Å. Diffracted neutrons are collected on a 80° (resolution 0.2°) 2θ range. Scans for combined analysis were operated from $\chi = 0$ to 90° (step 5°) using a fixed incidence angle ω of 25.11°. The average volume of the sample is 100 mm^3, and the measuring times were around 20 min per sample orientation.

Electrical resistivity measurements were performed by the standard dc four-probe technique from 5 to 400K at self field in a Physical Properties Measurement System (PPMS) from Quantum Design. The scanning electron microscopy (SEM) observations were made using a Zeiss Supra 55.

3. RESULTS AND DISCUSSION

The figure 1 illustrates three different $Bi_2Ca_2Co_{1.7}O_x$ grown bars according to their initial green shapes before the translating growth (50 mm/h). It evidences the potential of the technique to grow cylindrical rods with diameters of several millimeters.

Figure 1: Bi₂Ca₂Co₁.₇Oₓ grown rods depending of their initial green shapes before the translating growth (50 mm/h).

The figure 2 shows the transversal fracture along the bar axis. It can be clearly identified that the grains grow preferentially parallel to the cylindrical axis whereas an initial angle growth of around 30° is observed. In figure 2b, a close view of the fracture shows large platelike grains, with dimensions of several hundred of micrometers in the *ab* planes, and several micrometers in the *c* direction. As speculated from the microstructures, the texture (*i.e.* the c-axes of the misfit structure are preferentially aligned perpendicular to the rod axis) has been attested by neutron diffraction.

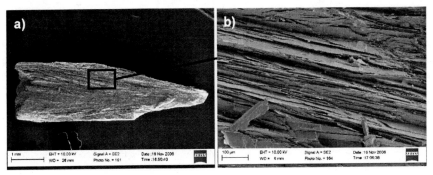

Figure 2: SEM micrographies of (a) a transversal fracture along the rod axis and (b) a close view showing the large platelike grains. Bi₂Ca₂Co₁.₇Oₓ compound.

The figure 3 presents the measured neutron diffraction pattern for all χ orientations of the sample. Firstly, it can be clearly seen that major peaks correspond to the misfit cobaltite Bi₂Ca₂Co₁.₇Oₓ whose the general monoclinic structure is described by two sublattices with their different *b* parameters: a=4.90Å b_1=4.73Å, b_2=2.80Å, c=14.66Å, β=93°49.[12] Secondly, this graph highlights without ambiguity the texture. In particular, we clearly observe the intensity decrease of the (*hk*0) peaks when χ increases and the appearance of the (00*l*) peaks when χ tends to 90°.

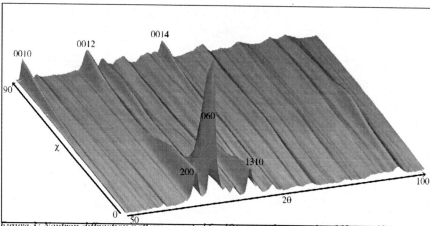

Figure 3: Neutron diffraction pattern operated for 19 χ-scans from χ = 0 to 90° (step 5°) using a fixed incidence angle ω of 25.11° ({006} Bragg position).

Based on a 3D structural model, reconstructed from the single crystal data, the whole diffraction pattern was refined. In figure 4, we can visually appreciate the agreement between the experimental (dots) and refined (lines) spectra for all the χ orientations. The refinement reliability is established by RP0, RP1 for the Orientation Distribution (OD) refinement, and R_w and RB factors for the Rietveld data, equal to 9.5%, 7.1 %, 9.6%, and 4.2%, respectively.

Figure 5 shows the inverse pole figures calculated for the z fiber direction, parallel to the translation axis. It shows a preferential orientation of the a and b-axis along the translation direction. Whereas we were attending a symetric planar texture, with an equivalent orientation of the [hk0] direction along the translation axis, the results show that the melting zone has induced a prenfrential growth of the platelike grains along some precise directions.

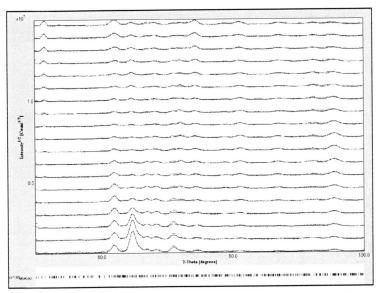

Figure 4: Experimental (dots) and calculated (lines) neutron diffraction patterns for various χ positions (0 to 90°).

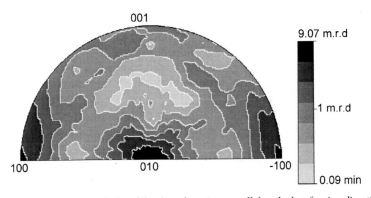

Figure 5: Inverse pole figures calculated for the z direction, parallel to the hot-forging direction.

In terms of thermoelectric properties, at that time, only electrical resistivity measurements have been performed. These ones are presented in figure 6. According to the literature,[13-14] the cylindrical rods, synthesized from the Bi$_2$Sr$_2$Co$_{1.8}$O$_x$ composition, exhibit a semiconductor/metal transition around 100-150K, whatever the translation speed growth is. On the opposite, the Bi$_2$Ca$_2$Co$_{1.7}$O$_x$ and Bi$_2$(Ca,Sr)$_2$Co$_{1.75}$O$_x$ compositions exhibit a semiconducting behavior on the full range of temperature, according to our last study on the BiCaCoO system.[15] As a general trend, *i.e.* for the three compositions, the translation speed has a significant influence on the transport properties. The decreasing up to 15 mm/h allows a reduction of the resistivity by a factor 1.5-2, whatever the composition is. This indicates that the crystal growth has an impact on the texturation and consequently on the transport behavior. Actually, we have not all the data to argue our hypothesis, but we strongly believe that the texture strength, the density and the crystallite size are the major factors of the electrical resistivity variations.

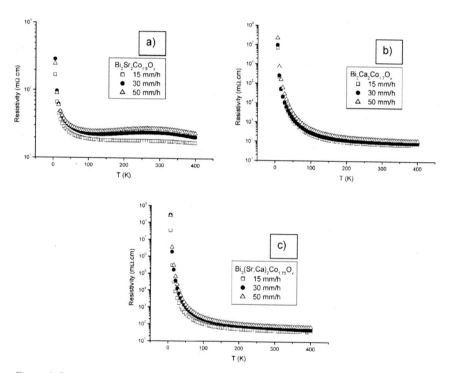

Figure 6: Resistivity versus temperature curves, for the (a) Bi$_2$Ca$_2$Co$_{1.7}$O$_x$, (b) Bi$_2$Sr$_2$Co$_{1.8}$O$_x$ and (c) Bi$_2$(Ca,Sr)$_2$Co$_{1.75}$O$_x$ compositions, depending of the LFZ translation speed.

Finally, the figure 7 presents the resistivity versus temperature curves of the three compositions for a fixed LFZ translation speed of 15 mm/h. The magnitude of the resistivity is

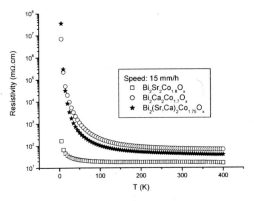

Figure 7: Resistivity versus temperature curves, for the three compositions. LFZ translation speed = 15 mm/h.

reduced by modifying the alkaline earth cations, which indicates and confirms[16] that the increase of the ionic radius of the alkaline earth cations induces an increase of the misfit ratio, thereby resulting in increasing the Co ions valence in the CoO$_2$ layer and consequently increasing the carrier concentration. For confirming this hypothesis, thermopower measurements will be carried out soon.

4. CONCLUSIONS

Preliminary results highlight the reliability and effectiveness of the LFZ method for texturing Bi-based misfit cobaltite rods. Neutron diffraction evidenced the texture developed during the melting growth. The electrical resistivity measurements of the three studied compounds, *i.e.* Bi$_2$Ca$_2$Co$_{1.7}$O$_x$, Bi$_2$Sr$_2$Co$_{1.8}$O$_x$, and Bi$_2$(Ca,Sr)$_2$Co$_{1.75}$O$_x$, depending of the translation speed, are in agreement with the results reported in the literature. Other physical properties characterizations (thermopower, thermal conductivity) are now under investigation in order to check the role of the composition and microstructure on the transport properties.

REFERENCES
1) I. Terasaki, Y. Sasago, and K. Uchinokura, Phys. Rev. B 1997, 56, R12685.
2) R. Funahashi, I. Matsubara, H. Ikuta, T. Takeuchi, U. Mizutani, and S. Sodeoka, Jpn. J. Appl. Phys. 2000, 39 L1127.
3) A.C. Masset, C. Michel, A. Maignan, M. Hervieu, O. Toulemonde, F. Studer, B. Raveau, and J. Hejtmanek, Phys. Rev. B 2000, 62, 166.
4) H. Leligny, D. Grebille, O. Perez, A. C. Masset, M. Hervieu, and B. Raveau, Acta Cryst. B 2000, 56, 173.
5) A. Maignan, S. Hébert, M. Hervieu, C. Michel, D. Pelloquin, and D. Khomskii, J. Phys.: Condens. Matter. 2003, 15, 2711.
6) H. Itahara, C. Xia, J. Sugiyama, and T. Tani, Chem. Mater., 2004, 14, 61.

7) E. Guilmeau, M. Mikami, R. Funahashi, and D. Chateigner, J. Mater. Res. 2005, 20, 1002.

8.- G. F. de la Fuente, J. C. Diez, L. A. Angurel, J. I. Peña, A. Sotelo, R. Navarro, Adv. Mater., 1995, 7, 853.

9.- Y. Huang, G. F. de la Fuente, A. Sotelo, A. Badía, F. Lera, R. Navarro, C. Rillo, R. Ibáñez, D. Beltran, F. Sapiña, A. Beltran, Physica C, 1991, 185, 2401.

10) L. Lutterotti, S. Matthies, H. R. Wenk, Proceedings of the 12th International Conference on Textures of Materials, Vol. 2, edited by J. A. Szpunar, 1999, 1599-1604. Montreal: NRC Research Press., Freeware available at: http://www.ing.unitn.it/~luttero/maud/

11) E. Guilmeau, D. Chateigner, J. Noudem, R. Funahashi, S. Horii, and B. Ouladdiaf, J. Appl. Cryst. 2005, 38, 199.

12) D. Grebille, H. Muguerra, E. Guilmeau, H. Rousselière, and R. Cloots, Chem. Mater. (Submitted)

13) T. Fujii and I. Terasaki, cond.mat. 2002, 0210071, New thermoelectric Materials Workshop:Beyond Bismuth Telluride".

14) T. Yamamato, K. Uchinokura, and I. Tsukada, Phys. Rev. B 2002, 65, 184434.

15) E. Guilmeau, M. Pollet, D. Grebille, M. Hervieu, H. Muguerra, R. Cloots, M. Mikami, and R. Funahashi, Inorg. Chem. 46 (2007) 2124.

16) H. Itahara, C. Xia, J. Sugiyama, and T. Tani, Journal of Materials Chemistry, 2004, 14, 61

GROWTH AND CHARACTERIZATION OF GERMANIUM-BASED TYPE I CLATHRATE THIN FILMS DEPOSITED BY PULSED LASER ABLATION

Robert Hyde, Matt Beekman, George S. Nolas, Pritish Mukherjee, and Sarath Witanachchi
Laboratory for Advanced Material Science and Technology (LAMSAT), Department of Physics,
University of South Florida
4202 East Fowler Avenue
Tampa, Florida 33620

ASTRACT
    Thin films of type I clathrate material, $Ba_8Ga_{16}Ge_{30}$, have been grown using the pulsed laser ablation technique. These materials hold the potential for thermoelectric applications. Films have been deposited on a variety of substrates including silicon, quartz, sapphire, glass substrates, and YSZ cubic zirconium. Optimal crystalline films were obtained at a growth temperature of 400°C and laser fluences at 3 J/cm² and above. The optimum laser fluence to produce stoichiometric films with the lowest particle density has been determined. The low ablation threshold for the UV wavelength leads to the particulate ejection during laser-target interaction. Excitation of the UV laser generated plasma by a second pulsed IR laser further reduced particulates and produced broader expansion profiles leading to large area uniform films.

INTRODUCTION

    The thermoelectric effect, or Peltier–Seebeck effect, is the conversion of temperature differentials to electric voltage and vice versa [1]. As Seebeck observed in 1821, when a thermal gradient is maintained along the length of a material, a thermoelectric field that is proportional to the temperature gradient is induced. The electric potential produced by a temperature difference is known as the Seebeck effect and the proportionality constant is called the Seebeck coefficient (S) where $S = \Delta V / \Delta T$ and is governed by the intrinsic properties of the material. Thermoelectrics are based on the 1834 discovery of the Peltier effect, by which DC current applied across two dissimilar materials causes a temperature differential. The Seebeck effect acts on a single conductor, whereas the Peltier effect is a typical junction phenomenon [2].

    In a thermoelectric material there are free carriers which carry both charge and heat. If a temperature gradient exists along a thermoelectric material, where one end is cold and the other is hot, the carriers at the hot end will move faster than those at the cold end. The faster hot carriers will diffuse further than the cold carriers and so there will be a net build up of carriers (higher density) at the cold end. The density gradient will cause the carriers to diffuse back to the hot end. In the steady state, the effect of the density gradient will exactly counteract the effect of the temperature gradient so there is no net flow of carriers. The buildup of charge at the cold end will also produce a repulsive electrostatic force (and therefore electric potential) to push the charges back to the hot end. If the free charges are positive (the material is p-type), positive charge will build up on the cold end which will have a positive potential (ΔV_p). Similarly, negative free charges (n-type material) will produce a negative potential at the cold end (ΔV_n) [3]. Therefore, a p-type and an n-type semiconductor can be connected to form an electrical series by thermally parallel device that would have a net potential difference of $|\Delta V_p| + |\Delta V_n|$. The Seebeck coefficient is very low for metals, being a few μV/K, while it is greater than 100 μV/K for semiconductors. A typical arrangement of a power-generation module is shown in Figure 1.

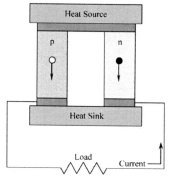

Figure 1. Typical thermoelectric device for power generation.

In order to convert thermal power into electrical power (IV) the following properties are desirable for the material; (a) low thermal conductivity, κ, to maintain the temperature gradient, (b) relatively high electrical conductivity, σ, to support a high current, and (c) a high voltage drop across the material per unit temperature, which is the Seebeck coefficient S. The combination of these requirements for a high efficiency thermoelectric device is represented by the figure-of-merit parameter (ZT) defined as $ZT = S^2\sigma T/\kappa$. Since metals and insulators have low values of σ/κ, good candidates for high ZT values are narrow band gap semiconductors [4].

Thermoelectric materials research over the last 30 years has obtained a figure-of-merit (ZT) of about 1. If a material system with a dual character can be constructed, where it behaves like a phonon glass towards thermal phonons to scatter them while behaving like a defect-free crystal towards electrons to maximize the electrical conduction, a so-called "phonon-glass electron-crystal" (PGEC) [5, 6]. With this, the research in thermoelectric materials is heading toward the discovery of an optimized PGEC material system where ZT values of 4 are possible [7]. The class of materials known as clathrates fulfills many of the PGEC requirements [8] and it was shown that these materials include potential candidates for thermoelectric applications, by Nolas *et al.* [9].

Clathrate compounds comprise a class of materials in which frameworks encapsulate loosely bonded atoms or molecules, commonly referred to as 'guests'. The type I clathrate structure is characterized by a framework typically composed of Group IV elements, which occupy sites at the vertices of face-sharing polyhedra, six tetrakaidecahedron and two pentagonal dodecahedron per cubic unit cell. Guest atoms in turn occupy the crystallographic sites found inside these polyhedra, and a general chemical formula for type I clathrates can be written as $A_8X_yY_{46-y}$, where A represents the guest atom, and X and Y represent the Group IV or substituting framework atoms [10]. The framework polyhedra that forms the type I clathrate structure is shown in Figure 2.

Type I clathrates show promising thermoelectric properties [9]. Some type I clathrates such as $Sr_8Ga_{16}Ge_{30}$ possess very low thermal conductivities, with temperature dependences similar to those of amorphous SiO_2 [7,11]. This has been explained by the unique interaction of the host framework heat-carrying phonons with the localized vibrational modes of the guests which may scatter these phonons [7, 12]. The loosely bound guests in these crystalline materials

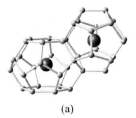

(a)

Figure 2. The framework polyhedra that forms the type I clathrate structure. The framework atoms are shown in light grey, while the guest atoms are shown in dark grey.

can undergo large, localized low-frequency vibrations, which can scatter the heat-carrying acoustic phonons resulting in thermal conductivities with magnitudes similar to amorphous materials. The low thermal conductivity combined with the relatively high power factors reported at high temperature for clathrates such as $Ba_8Ga_{16}Ge_{30}$ [13] are why these materials continue to be actively investigated in the field of thermoelectrics. The clathrate guest-host interaction continues to be of scientific interest and as a result of the interesting properties and technological promise of type I clathrates, bulk properties of these materials have been studied extensively using a wide range of experimental and theoretical techniques. However, to date no reports exist on the production and characterization of type I clathrate thin films [14]. In this paper we report the first results on the growth and structural characterization of germanium-based clathrate thin films of $Ba_8Ga_{16}Ge_{30}$ using pulsed laser deposition. One of the main advantages of laser ablation for thin film growth is the ability of this process to closely reproduce the target stoichiometry in the deposited film [15-17]. Therefore, laser ablation is uniquely suited for the growth of multi-component films from a single composite target such as $Ba_8Ga_{16}Ge_{30}$.

EXPERIMENTAL DETAILS

Type I clathrates used for laser ablation targets were prepared by mixing stoichiometric quantities of the high purity elements Ba (99%, Aldrich), Ga (99.9999%, Chameleon), and Ge (99.99%, Alfa Aesar). The mixtures were placed in pyrolytic boron nitride crucibles, and sealed in fused quartz ampoules under high purity nitrogen gas at a pressure of 2/3 atmosphere. The mixtures were heated at 1°C/min to form $Ba_8Ga_{16}Ge_{30}$, held at 1000°C for 24 hours, and then cooled at a rate of 2°C/min. Targets for ablation were produce by hot and cold pressing procedures to a compacted density of higher than 95% of the theoretical X-ray density, as analyzed by powder X-ray diffraction (XRD) [14].

The pulsed laser ablation system used for single laser and dual laser ablation film growth is shown in Figure 3. In the single laser ablation process excimer laser pulses with a wavelength of 248 nm and duration of 25 ns are focused onto the rotating target placed in a 10^{-7} torr vacuum. The laser-target interaction generates a plasma plume of the target species that expands into the vacuum. The plume is allowed to deposit on a heated substrate placed on-axis 6 cm from the target. For dual-laser ablation, a CO_2 laser with pulse duration of 250 ns and wavelength of 10.06 μm was focused to spatially and temporally overlap the excimer laser spot on the target surface. The delay between the two lasers was monitored and adjusted to obtain the minimum particulate generation. Reported films of $Ba_8Ga_{16}Ge_{30}$ were deposited on quartz substrates from room temperature to 600°C to an on-axis thickness of 6000 Å. The stoichiometry was studied by energy dispersive spectroscopy (EDS) analysis, the crystal structure characterization of the films was performed by X-ray diffraction (XRD) analysis, and the surface morphologies of the films were investigated by scanning electron microscopy (SEM).

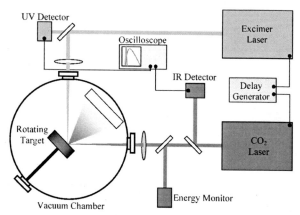

Figure 3. Diagram of the laser ablation system.

RESULTS AND DISCUSSION
    The on-axis film deposition rate per laser pulse as a function of the single excimer laser fluence is shown in Figure 4. The cut-off fluence is approximately 0.5 J/cm^2 below which no detectable deposition occurs. At low laser fluences the ion density and the energy of the evaporated species as well as the rate of film growth is significantly low.
    The elemental ratios of the films obtained by EDS analysis were similar for single laser

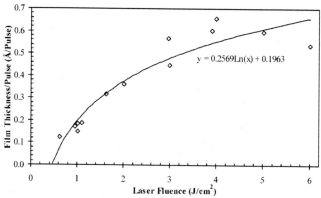

Figure 4. On-axis film deposition rate as a function of the incident single laser fluence.

ablation fluences in the range of 1 to 6 J/cm^2 and maintained the approximate stoichiometry of the target of Ba$_8$Ga$_{16}$Ge$_{30}$ as shown in Figure 5. However, the stoichiometries deviated at laser fluences lower than 1 J/cm^2.

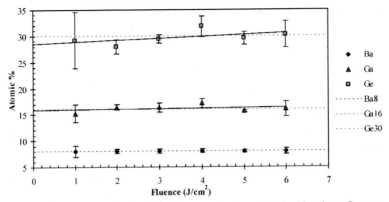

Figure 5. EDS atomic percent analysis of thin films as a function of the incident laser fluence.

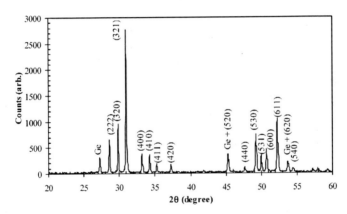

Figure 6. X-ray diffraction pattern of the Ba$_8$Ga$_{16}$Ge$_{30}$ target used for laser ablation.

Crystallinity of the films was determined by X-ray diffraction. The X-ray diffraction peaks produced by the target is shown in Figure 6. Figure 7 shows the x-ray diffraction patterns for films deposited in the temperature range of room temperature to 500 °C. The clathrate peak corresponding to the (321) orientation is clearly seen. A maximum intensity of the peak appearing at 2Θ equal to 31.2° (321) occurs at a substrate temperature of 400°C, as shown in Figure 8. Figure 9 shows the x-ray diffraction patterns for films deposited at a substrate temperature of 400 °C in the laser fluence range of 1 to 6 J/cm^2 as compared to the target pattern.

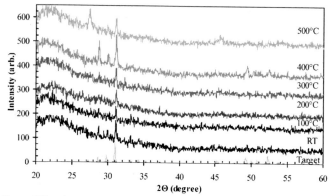

Figure 7. X-ray diffraction patterns of clathrate films on quartz substrates at temperatures from room temperature to 500°C.

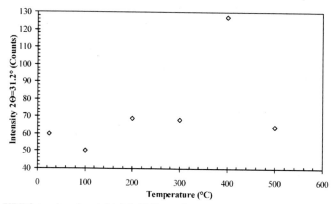

Figure 8. XRD intensity of peak 31.2 ° (321) as a function of temperature.

The intensity of the peak appearing at 2Θ equal to 31.2° (321) at the various laser fluences is shown in Figure 10. The intensity rises to a maximum and plateaus at approximately 3 J/cm^2.

With increasing laser fluence the ion density of the plasma increases leading to enhanced absorption of the laser radiation into the plasma. This increases the plasma temperature that causes the particulates in the plasma to re-evaporate. However, for very high fluences density of particles ejected increases as well. The surface morphology of $Ba_8Ga_{16}Ge_{30}$ films deposited with laser fluences from 1 J/cm^2 to 4.0 J/cm^2 have been studied by SEM. Particulate density was relatively high for both low and high fluences, as shown in Figure 11. Large molten droplets occur at higher laser fluences as is problematic with metallic sources. If the molten zone of the laser-target spot is further heated by pumping it with a CO_2 laser pulse, further reduction in particulate ejection can be expected. Figure 12 shows an example of the oscilloscope traces of

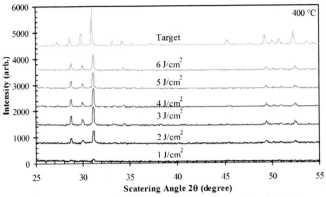

Figure 9. X-ray diffraction patterns of the clathrate films at laser fluences in the range of 1 to 6 J/cm^2.

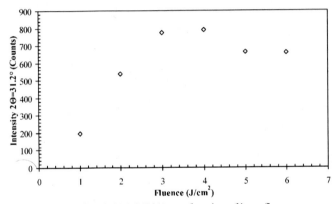

Figure 10. XRD intensity of peak 31.2 ° (321) as a function of laser fluence.

the laser pulse timing between the UV excimer laser and the IR CO_2 laser at a peak-to-peak delay of 100 ns (-22 ns from the onset of the CO_2 laser pulse to the onset of the excimer laser pulse) where the IR pulse arrives slightly before the excimer pulse. Figure 13 shows SEM images of a single laser ablated film and three films utilizing the dual laser technique at various laser delays; 146 ns peak-to-peak (2 ns from onset), 100 ns peak-to-peak (-22 ns from onset), and -23 ns peak-to-peak (-140 ns from onset). The excimer laser was operated at a fluence of 1 J/cm^2 and the CO_2 laser at a fluence of 0.4 J/cm^2. The minimum particulate generation was observed while using a 100 ns delay. As seen in Figure 13(c) that the majority of the large particles have been eliminated by maintaining a melt-zone on the target surface produced by the dual laser process. Optimization of the technique is being investigated to further reduce the particulates.

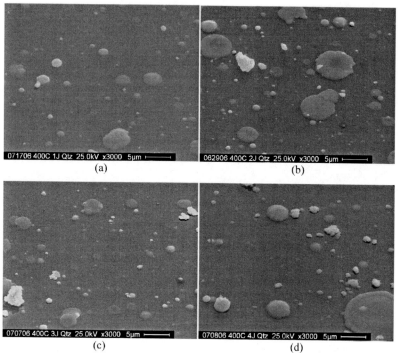

Figure 11. SEM images of clathrate films deposited with single laser ablation fluences of (a) 1 J/cm^2 (b) 2 J/cm^2 (c) 3 J/cm^2 and (d) 4 J/cm^2.

SUMMARY

We have successfully demonstrated the use of pulsed laser ablation techniques for the stoichiometric growth of crystalline thin films of the type I clathrate $Ba_8Ga_{16}Ge_{30}$. The ablation threshold for the material at UV wavelengths appears to be low, approximately 0.5 J/cm^2. Stoichiometry of the films deposited near this threshold deviated from the required 8:16:30 ratio. Partial evaporation of the species at the target at low laser fluences may be responsible for this effect. However, for laser fluences above 1 J/cm^2 films retained the proper stoichiometry. With increasing fluence the density of particles deposited on the film also increased. We have utilized a dual-laser approach by coupling the energy of a CO_2 laser into the laser-target molten-zone at the optimum inter-pulse delay to minimize the ejection of particulates from the target. The experiments showed that for optimum delay between the onset of the excimer and CO_2 laser pulses to produce films with low particle density is approximately 22 ns (100 ns peak-to-peak).

ACKNOWLEDGEMENTS

This project is partially supported by the US Department of Energy, under Grant No. DE-FG02-04ER46145 and by the National Science Foundation, under Grant No. DMI-0217939.

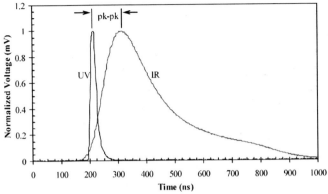

Figure 12. An example of oscilloscope traces of the excimer pulse (UV) and the CO_2 laser pulse (IR) with a 100 ns peak-to-peak delay.

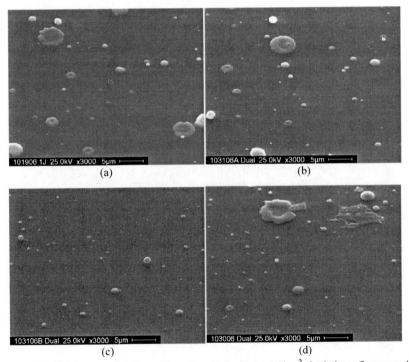

Figure 13. SEM images of clathrate films deposited with (a) 1 J/cm^2 single laser fluence and dual laser peak-to-peak delays of (b) 146 ns (c) 100 ns and (d) -23 ns.

REFERENCES

[1] R. M. Besançon (1985). *The Encyclopedia of Physics, Third Edition*. Van Nostrand Reinhold Co. (1985).

[2] Melcor Corp., *Thermoelectric Handbook*, Sept., (1995).

[3] *CRC Handbook of Thermoelectrics, Introduction*, D.M. Rowe (ed.), CRC Press, (1995).

[4] G. S. Nolas, G. A. Slack, and S. B. Schujman. "Semiconductor Clathrates: A Phonon-Glass Electron-Crystal Material with Potential for Thermoelectric Applications" in *Recent Trends in Thermoelectric Materials Research I*, T. M. Tritt (ed.) *Semiconductors and Semimentals* V. 69 (Academic Press 2000).

[5] T. M. Tritt (ed.) *Semiconductors and Semimetals* V. 69, 70, and 71 (Academic Press, 2001)

[6] G. A. Slack "New Materials and Performance Limits for Thermoelectric Cooling" CRC Handbook of Thermoelectrics, ed. D. M. Rowe 407-440 (CRC Press. 1995).

[7] J. P. Heremans, C. M. Thrush, and D. T. Morelli, "Thermopower Enhancement in Lead Telluride Nanostructures", Phys. Rev. B 70, 115334 (2004).

[8] G. A. Slack, Mater. Res. Soc. Symp. Proc. **478** 47 (1997).

[9] G. S. Nolas, J. L. Cohn G. A. Slack, and S. B. Schujman. Appl.Phys. Lett. 73. 187 (1998).

[10] G. S. Nolas, G. A. Slack, and S. B. Schujman. "Semiconductor Clathrates: A Phonon-Glass Electron-Crystal Material with Potential for Thermoelectric Applications" in *Recent Trends in Thermoelectric Materials Research I*, T. M. Tritt (ed.) *Semiconductors and Semimentals* V. 69 (Academic Press 2000).

[11] J. L. Cohn G. S. Nolas, V. Fessatidis, T. H. Metcalf, and G. A. Slack. Phys. Rev. Lett. **82.** 779 (1999).

[12] Nolas, G.S., Weakley, T.J.R., Cohn, J.L., and Sharma, R., "Structural properties and thermal conductivity of crystalline clathrates," *Phys. Rev. B* **61**, 3845 (2000).

[13] Saramat, A. et al., "Large thermoelectric figure of merit at high temperature in Czochralski-grown clathrate $Ba_8Ga_{16}Ge_{30}$," *J. Appl. Phys.* **99**, 023708 (2006).

[14] S. Witanachchi, R. Hyde, H. S. Nagaraja, M. Beekman, G. S. Nolas, and P. Mukherjee, "Growth and Characterization of Germanium-based Type I Clathrate Thin Films Deposited by Pulsed Laser Ablation" Mater. Res. Soc. Symp. Proc. Vol. 886 Materials Research Society 401-406 (2006).

[15] Witanachchi, S., Ahmed, K., Sakthivel, P., and Mukherjee, P., "Dual-laser ablation for particulate-free film growth *Appl. Phys. Lett.* 66, 1469 (1995).

[16] Mukherjee, P., Cuff, J.B., and Witanachchi, S., "Plume expansion and stoichiometry in the growth of multi-component thin films using dual-laser ablation," *Appl. Surface Sci.* **127-129**, 620 (1998).

[17] Mukherjee, P., Chen, S., Cuff, J.B., Sakthivel, P., Witanachchi, S., "Evidence for the physical basis and universality of the elimination of particulates using dual-laser ablation. I. Dynamic time-resolved target melt studies, and film growth of Y2O3 and ZnO," *J. Appl. Phys.* **91**, 1828 (2002).

SYNTHESIS AND CHARACTERIZATION OF CHALCOGENIDE NANOCOMPOSITES

J. Martin and G. S. Nolas
Department of Physics, University of South Florida, Tampa, Florida 33620

ABSTRACT

Lead Telluride (PbTe) nanocomposites were prepared by densifying 100 – 150 nm PbTe nanocrystal powders synthesized from a low temperature alkaline aqueous solution reaction in a high yield (> 2 grams per batch). Densification using Spark Plasma Sintering (SPS) successfully integrated nanostructure dispersions within a bulk matrix. We report the synthesis and low temperature transport measurements on novel chalcogenide nanocomposites, including resistivity, Seebeck coefficient, and thermal conductivity.

INTRODUCTION

Thermoelectric effects couple thermal and electric currents, allowing for the solid-state inter-conversion of heat and electricity. The thermoelectric figure of merit, $ZT = S^2 \sigma T / \kappa$, defines the effectiveness of different thermoelectric materials, where S is the Seebeck coefficient, σ is the electrical conductivity, T is the absolute temperature, and κ is the total thermal conductivity. Recent progress in a number of higher efficiency thermoelectric materials (room temperature ZT > 1) can be attributed to the nanoscale enhancement of thermoelectric properties. These new materials demonstrate increased Seebeck coefficient and decreased thermal conductivity due to the phenomenological properties of nanometer length scales, introducing both quantum confinement and interfacial scattering effects.

One consequence of nanostructure is the increase of interfaces. These interfaces serve to scatter phonons more effectively than electrons and reduce the thermal conductivity. Additionally, the presence of interfacial energy barriers filters the carrier energy traversing the interface, restricting those energies that limit the mean carrier energy.[1] This increases the Seebeck coefficient, as its value depends on the mean carrier energies relative to those at the Fermi level. Experimental evidence in expitaxial films of n- and p-type lead chalcogenides verifies the enhancement potential of this mechanism.[2]

The primary research into nanostructured enhancement of thermoelectric properties remains limited to thin films, heterostructures, and nanowires. For example, p-type Bi_2Te_3/Sb_2Te_3 10 angstrom/50 angstrom supperlattice structures demonstrated a room temperature ZT = 2.4. The thermal conductivity in this system was reduced by a factor of 2 compared to other Bi_2Te_3 alloys.[3] Harman and co-workers reported a room temperature ZT = 1.6 in PbTe/PbTeSe quantum-dot superlattices (QDSLs)[4] that contain PbSe nanodots imbedded in a PbTe matrix. Kong and co-workers reported an enhancement in Si/Ge supperlattices.[5] Recently, Heremans and co-workers reported an increased Seebeck coefficient for PbTe with the inclusion of Pb precipitate nanostructures.[6]

One method to incorporate nanoscale dimensions into bulk materials is by ball-milling.[7,8] This procedure rapidly grinds powders to sub-micron dimensions. Ball-milled Si-Ge nanocomposites demonstrated an increased Seebeck coefficient and a reduced thermal conductivity.[9] Although the electrical conductivity also increased, the overall thermoelectric performance of the material was enhanced. Ball-milled PbTe materials have also demonstrated thermoelectric enhancement.[10] However, the syntheses discussed leave suspect a direct correlation between nanostructure and thermoelectric enhancement, due to unaccounted lattice strain affects. This stresses the need to directly synthesize nanocrystals and incorporate them into bulk structures.

SYNTHESIS AND CHARACTERIZATION

Lead telluride nanocrystals were synthesized from the low temperature reaction of a selenium or tellurium alkaline aqueous solution and a lead acetate trihydrate solution.[11] Several experiments were performed to optimize product yield by varying both the alkaline and precursor lead acetate trihydrate concentrations. To prepare PbTe nanocrystals, the precursor solutions were prepared separately at 90° C by dissolving elemental Te in a 20 M KOH aqueous solution and by dissolving $Pb(CH_3COO)_2·3H_2O$ in distilled water. After ~ 30 minutes, the lead acetate trihydrate solution was dripped into the rapidly stirring deep purple alkaline solution to immediately form PbTe nanocrystals. After 5 minutes, the reaction mixture was removed from the heat source and dilute HNO_3 was added to flocculate the nanocrystals. The solution was removed from the grayish-black precipitates that were subsequently washed 4 times with the dilute nitric acid solution, removing lead hydroxide impurities. Excess lead acetate trihydrate in the reaction favors the formation of easily removable impurities. The precipitates were then washed 4 times with distilled water and dried overnight in a fume hood. XRD analysis confirmed nanocrystal phase purity. This procedure reproducibly synthesizes 100 – 150 nm spherical PbTe nanocrystals, confirmed by TEM (Figure 1), with a high yield of over 2 grams per batch. High yields of nanocrystals are required to synthesize practical nanocomposite materials.

PbTe nanocrystals were subjected to Spark Plasma Sintering (SPS) to achieve ~ 95 % theoretical density. In the SPS procedure, a pulsed DC current conducts through both the graphite die and the sample under high pressure to minimize nanostructure deterioration. Prior to SPS, five batches of PbTe nanopowders were mixed together in a glass vial for each sample. The final densities are 7.67 g/cm^3 (sample PbTe1) and 7.75 g/cm^3 (sample PbTe2). Densifying solely the nanocrystals results in a uniform dispersion of non-conglomerated nanostructure within a bulk matrix.

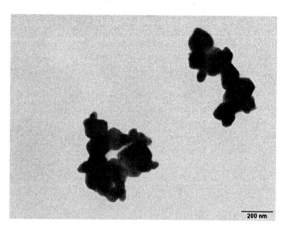

Figure 1. TEM image of PbTe nanocrystals after precipitation.

Figure 2. SEM image indicating the nanostructure of PbTe1.

SEM (Hitachi S-800) images of a PbTe1 fracture surface indicate the preservation of nanostructure following the SPS procedure with grain diameters ranging from 100 nm to over one micron (Figure 2). Figure 3 shows the standard X-ray diffraction scans for the two PbTe nanocomposites post SPS. All samples exhibit peaks characteristic of PbTe. The successive spectra are normalized and shifted in intensity for clarity. While the PbTe nanopowder spectra indicates phase purity, the sintered nanocomposites exhibit a PbTeO$_3$ impurity phase.

Densified nanocomposites of PbTe were cut into 1 x 2 x 5 mm parallelopipeds for transport property measurements. Four-probe resistivity, Seebeck coefficient, and steady-state thermal conductivity were measured on the same sample from 300 K – 12 K. Table 1 summarizes the room temperature data. Both samples are p-type with large room temperature Seebeck coefficients of ~ 330 μV/K (Figure 4). The temperature dependence and the magnitude of the Seebeck share consistency throughout the measured temperature range. Transport measurements also demonstrate a correlation between resistivity and porosity (Figure 5). An increase of only 1.5 % theoretical density between PbTe1 and PbTe2 decreased the room temperature resistivity by a factor of 2. Since porosity has no affect on the Seebeck coefficient[12], increasingly denser nanocomposites should facilitate optimal thermoelectric performance. The thermal conductivity values also correlate to the difference in sample density. As shown in Table 1, PbTe1 demonstrates a slightly lower thermal conductivity than PbTe2 due to the increased porosity.[13] The room temperature ZT for PbTe2 is ~ 0.1 without optimization or doping.

Table 1. Percent theoretical density, resistivity, Seebeck coefficient, and total thermal conductivity at 300 K for the PbTe1 and PbTe2 nanocomposites.

Sample	Density (%)	ρ (mOhm-cm)	S (μV/K)	κ (W-m^{-1}K^{-1})
PbTe1	94	24.9	328	2.2
PbTe2	95	12.6	324	2.5

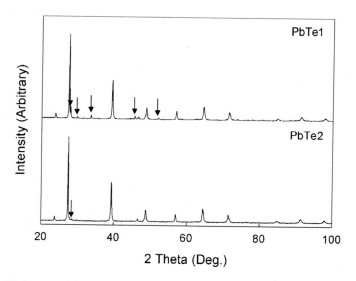

Figure 3. XRD spectra for the two PbTe nanocomposites post SPS procedure. Arrows indicate PbTeO$_3$ impurity.

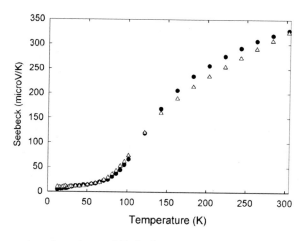

Figure 4. Temperature dependence of Seebeck coefficient for the two PbTe nanocomposites. Closed circles identify PbTe1 and open triangles identify PbTe2.

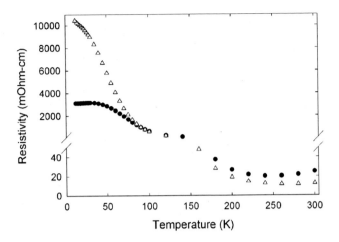

Figure 5. Temperature dependence of resistivity for the two PbTe nanocomposites. Closed circles identify PbTe1 and open triangles identify PbTe2.

CONCLUSIONS

Dense PbTe nanocomposites were prepared by densifying PbTe nanocrystal powders synthesized from high yield (> 2 grams per batch) alkaline aqueous solution reactions. These nanocomposites dimensionally integrate nanostructure within a bulk matrix without conglomeration. Transport measurements indicate a strong correlation between resistivity and porosity, suggesting increasingly denser nanocomposites might further optimize the transport properties for thermoelectric properties. To our knowledge, these samples represent the first preparation of dense nanocomposites from solution-phase synthesized nanocrystals, including transport measurements.

The authors acknowledge support by GM and DOE under corporate agreement DE-FC26–04NT42278.

REFERENCES

[1] B. Y. Moizhes and V. A. Nemchinsky, *In Proceeding for the 11th International Conference on Thermoelectrics*, Institute of Electrical and Electronics Engineers, Inc., 1992.
[2] Y. I. Ravich, In *CRC Handbook of Thermoelectrics*, edited by D. M. Rowe, pages 67-73, CRC Press New York, 1995.
[3] R. Venkatasubramanian, E. Siivola, T. Colpitts, B. O'Quinn, *Nature* 413, 597 (2001).
[4] T. C. Harman, P. J. Taylor, M. P. Walsh, and B. E. LaForge, *Science* 297, 2229 (2002).
[5] T. Kong, S. B. Cronin, M. S. Dresselhaus, *Applied Physics Letters* 77, 1490 (2000).
[6] J.P. Heremans, C.M. Thrush and D.T. Morelli, *J. Appl. Phys.* 98, 063703 (2005).
[7] K. Kishimoto and T. Koyanagi, *J. Appl. Phys.* 92, 2544 (2002).

[8] P.C. Znai, W.Y. Zhao, Y. Li, X.F. Tang, Z.J. Zhang and M. Nino, *Appl. Phys. Lett.* **89**, 052111 (2006).

[9] M. S. Dresselhaus, G. Chen, M. Y. Tang, R. G. Yang, H. Lee, D. Z. Wang, Z. F. Ren, J. P. Fleurial and P. Gogna, *Proc. Mater. Res. Soc.* **886** 3 (2006).

[10] J. P. Heremans, C. M. Thrush, and D. T. Morelli, Thermopower Enhancement in PbTe nanostructures, *Physical Review B* **70**, 115334 (2004).

[11] W. Zhang, L. Zhang, Y. Cheng, Z. Hui, X. Zhang, Y. Xie, and Y. Qian, *Materials Research Bulletin* **35**, 2009 (2000).

[12] L. Yang, J. S. Wu and L. T. Zhang, *J. Alloys and Compounds* **364**, 83 (2004).

[13] I.Sumirat, Y. Ando, S. Shimamura, *J. Porous Mater.* **13**, 439 (2006).

ANOMALOUS THERMAL CONDUCTIVITY OBSERVED IN THE $Na_{1-x}Mg_xV_2O_5$ SINGLE CRYSTALS

Z. Su, J. He, T. M. Tritt

Department of Physics and Astronomy
Clemson University
Clemson, SC 29634, USA

ABSTRACT

Single crystalline specimens of $Na_{1-x}Mg_xV_2O_5$ (nominally, x=0.00-0.35) have been grown using a self-flux technique and characterized by specific heat and thermal conductivity measurements. A novel thermal conductivity measurement system called the parallel thermal conductance method has been developed in our lab, which allows us to measure the thermal conductivity of these needle-like single crystals along their elongated b axis from 10 K up to 310 K. We observed a pronounced "dip" in the thermal conductivity at T_c, where a phase transition occurs. A close relation has been established between the Na/Mg concentration, the transition temperature T_c, the position and amplitude of the thermal conductivity dip and specific heat anomaly. The underlying mechanisms have been discussed in light of the disorder nature of this low-dimensional compound.

INTRODUCTION

The discovery of a phase transition at 34 K in NaV_2O_5 crystals via magnetic susceptibility χ measurements of NaV_2O_5 [1] has attracted a lot of attention, and subsequent investigations show that this phase transition can also be observed in thermal expansion α, [2] specific heat C_p, [2,3] and thermal conductivity κ [4] measurements. Based on the data in χ, it was at first proposed that the phase transition is a spin-Peierls transition. [1] However, this is not completely consistent with the phenomena observed in C_p [3] and κ, [4] thus making the origins of the phase transition still controversial. Spin-lattice coupling and/or charge ordering may also contribute to the anomaly, therefore it is more likely to be a cooperative phase transition.

It has been shown that the shape of the peak and position of the anomaly are insensitive to the magnetic field but can be affected by Na vacancies inside the samples. [3] Therefore, in our work we chose to dope NaV_2O_5 with Mg, which can supposedly occupy Na vacancies and also introduces one more electron to affect the valence on the V atoms and thus change the V^{4+} to V^{5+} ratio in the lattice.

EXPERIMENTS

The $Na_{1-x}Mg_xV_2O_5$ single crystals were grown using a self-flux method with a nominal doping ratio x, ranging from 0.00- 0.35. The $NaVO_3$ flux was first prepared by heating stoichiometric amounts of Na_2CO_3 (99.9997%), Mg_2CO_3 (99.99%), and V_2O_5 (99.995%) at 800 °C for 2 hours. Further VO_2 (99%) was added and the admixture was annealed at 725 °C and subsequently cooled down to room temperature at a rate of 1-2 °C/hr in an evacuated silica ampoule. After dissolving the $NaVO_3$ flux with hot water, the elongated platelets of Na_1.

$_xMg_xV_2O_5$ single crystals with typical dimensions of $0.2 \times 2.0 \times 0.5$ mm^3 were obtained. It was found the elongated direction was along b-axis.

For a traditional steady thermal conductivity measurement, such small samples are too brittle to support the thermocouple and the heater attached to the sample. Meanwhile, the heat leak through attached thermocouple and other media would be comparatively large because of the low thermal conductance of the sample. To overcome the difficulties, we developed a method called parallel thermal conductance technique, or PTC, in our lab, which is still a classical steady state method. [5] However, in the PTC technique, the sample stage's thermal conductance, or base line, is measured before a sample is installed. By subtracting the base line from the total thermal conductance of the sample and sample stage, the thermal conductance of the sample is acquired. Figure 1 shows the diagram of one PTC sample stage and a photograph of a PTC sample puck with two sample stages on it. Two samples can be measured simultaneously on one puck with two sample stages.

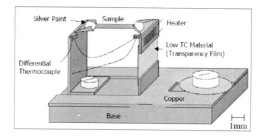 

Figure 1. PTC sample stage (left) and PTC sample puck with two sample stages (right)

The specific heat measurements on our samples were performed from 1.8~ 310 K on Quantum Design PPMS Model 6000 in our laboratory.

RESULTS AND DISCUSSION
Specific heat

We measured the specific heat of NaV_2O_5 samples with and without Mg doping, and the results are shown in Figure 2. Without Mg doping, NaV_2O_5 sample (x=0) has a "hump" at about 33 K. Schnelle et al. reported in their paper that the Na vacancies in $Na_{1-x}V_2O_5$ could affect the peak position and shape of the phase transition in the specific heat measurement and 5% Na vacancies could significantly suppress the peak.[3] By comparing our results with those on Schnelle's samples that contained Na vacancies, we may assume that our undoped NaV_2O_5 single crystal samples contain roughly 4% Na vacancies. Then a 5% Mg (x= 0.05) doping can dramatically change that hump into a steep and narrow peak without obvious peak position shifting, and the peak height is roughly the same as Schnelle's sample with 0% Na vacancies. However, as Mg doping ratio increases to 10% (x= 0.10) and 15% (x= 0.15), the peaks become lower and broader, and a shifting of peak positions to low temperature is observed. For 20% (x= 0.20) and 25% (x= 0.25) Mg-doped samples, this anomaly becomes a small hump that can be barely seen with systematic tendency of peak position's shifting to low temperature. This anomaly is finally smeared out as Mg doping ratio is increased to 30% (x= 0.30) and 35% (x= 0.35).

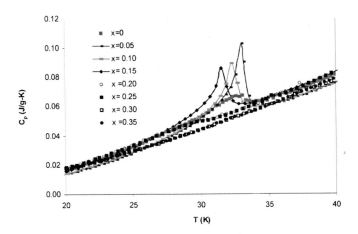

Figure 2. The specific heat, C$_p$, of Na$_{1-x}$Mg$_x$V$_2$O$_5$ (x= 0.00- 0.35)

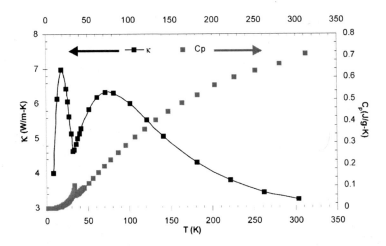

Figure 3. Anomalies of Na$_{1-x}$Mg$_x$V$_2$O$_5$ observed in thermal conductivity κ and specific heat C$_p$ measurements.

The effect of the 5% Mg doping on our NaV_2O_5 samples may be considered as an implication that Na vacancies in the lattice are occupied by Mg ions after doping. The effects of doping ratios higher than 5% on the samples could be explained as more distortions and mismatches introduced into the lattice by excess Mg ions.

Thermal Conductivity

By means of the PTC technique mentioned above, we have observed that, $Na_{1-x}Mg_xV_2O_5$ also has a dip in the thermal conductivity, κ, at about 32 K, where the peak in specific heat, C_p, occurs. These results are shown in Figure 3. This result confirms Vasil'ev's data acquired from undoped NaV_2O_5 single crystal samples. [4]

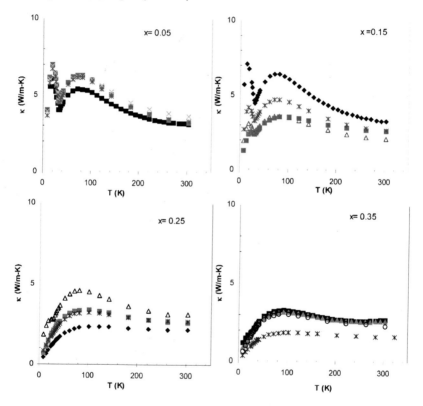

Figure 4. The thermal conductivity, κ, of $Na_{1-x}Mg_xV_2O_5$ samples categorized in different nominal Mg doping ratios

The C_p measurement has revealed that different nominal Mg doping ratios can change the peak's shape and position on temperature axis. In the thermal conductivity measurement, the sample's dimensions, can often have significant uncertainty, especially for samples that have irregular cross sections like $Na_{1-x}Mg_xV_2O_5$. This is in contrast to a C_p measurement, where the sample's mass can be weighed with rather high precision Therefore, when we study the relationship between the dip in κ and Mg doping ratios, the dip's relative depth and the temperature where it reaches the minimum, are of our highest interest. Furthermore, these quantities are independent of sample dimension measurements and its inherent uncertainty. Another issue that we need to take into consideration is the non-uniformity among samples with the same nominal Mg doping ratio. In order to address this issue, four samples of each nominal doping ratio were selected to measure their thermal conductivity using the PTC technique and the results are shown in Figure 4.

In summary, very deep "dips" in the thermal conductivity can be easily observed in all 5% (x= 0.05) Mg-doped samples. For x= 0.15, some samples' dips in κ begin to become more shallow. For x= 0.25, only one sample's anomaly can be barely described as a dip and the other three's are almost smeared out. As the doping ratio is increased to x= 0.35, the dips in κ of all samples are almost completely smeared out. Within one doping ratio category, the deeper the dip in κ, the lower the temperature where its corresponding minimum occurs.

CONCLUSION

Our novel PTC technique has allowed us to measure this complimentary thermal transport measurement of the thermal conductivity in $Na_{1-x}Mg_xV_2O_5$ single crystals for the first time in order to be compared to results on the heat capacity concerning the low temperature phase transition. The $Na_{1-x}Mg_xV_2O_5$ single crystals exhibit a peak in C_p measurement at 33K and a dip in κ at 32K, both of which correspond to a phase transition previously reported by others. After doping, this phase transition can be either enhanced by Mg ions' occupation of Na vacancies, or weakened if samples are excessively doped. Doping can also shift the critical temperature of this phase transition. Our future work will include the quantitative analysis of actual Mg doping ratios in the samples to set up a more accurate relation between the doping and the phase transition.

ACKNOWLEDGEMENT

This research is funded by Department of Energy Implementation Program and SC EPSCoR/ Clemson University Cost Share. We also appreciate the Oak Ridge National Laboratory where the samples were synthesized.

REFERENCES

1 M. Isobe and Y. Ueda, Journal of the Physical Society of Japan 65, 1178 (1996).
2. M. Köppen, D, Pankert, R. Hauptmann et al., Physical Review B 57, 8466 (1998).
3. W. Schnelle, Yu. Grin, and R. K. Kremer, Physical Review B 59, 73 (1999).
4. A. N. Vasil'ev, V. V. Pryadun, D. I. Khomskii et al., Physical Review Letters 81, 1949 (1998).
5. B. M. Zawilski, R. T. Littleton IV, and T. M. Tritt, Review of Scientific Instruments 72, 1770 (2001).

PHYSICAL PROPERTIES OF HOT-PRESSED $K_8Ge_{44}\square_2$

M. Beekman and G.S. Nolas
Department of Physics, University of South Florida
Tampa, FL 33620, USA

ABSTRACT

We report on the preparation and physical properties of polycrystalline type I Ge clathrate $K_8Ge_{44}\square_2$ ($\square$ = framework vacancy), with K encapsulated inside the Ge framework polyhedra. Temperature dependent resistivity data indicate our $K_8Ge_{44}\square_2$ specimen behaves as a poor metal, with a modest room temperature Seebeck coefficient of ~ -80 μV/K. The thermal conductivity of the $K_8Ge_{44}\square_2$ specimen was found to be very low, comparable in magnitude at room temperature to other type I clathrates. Thermal analysis reveals this compound to be a meta-stable phase. By comparison with previous work, our results indicate that the properties of this material are sensitive to the synthesis conditions. The ideal Zintl-Klemm concepts, while applicable qualitatively, may not strictly hold for this compound.

INTRODUCTION

Inorganic clathrate materials are characterized by their open-framework crystal structure that allows for the encapsulation of guest atoms inside atomic polyhedra formed by a covalently bonded framework. The most conspicuous aspect of these materials is the relationship between the guest atoms and the host framework, a relationship that has important consequences for the properties of these materials. There are several known structural types of inorganic clathrates.[1] The bulk properties of type I clathrates have been extensively studied, in large part due to the promise these materials hold for thermoelectric applications,[2,3] but also due to the superconducting[4] and magnetic properties[5,6] that some clathrates possess. Still, the wide compositional variety possible for type I clathrates allows for the study of new variants with this structure type.

The type I clathrates crystallize with the space group $Pm\overline{3}n$; a schematic of the crystal structure is shown in Fig. 1. As seen from Fig. 1b, one may view the structure as composed of face-sharing polyhedra, which form the covalently bonded framework. The framework is typically formed by Group IV elements (Si, Ge, or Sn), or substituting species such as Group III elements or transition metals. The guest atoms then occupy the voids formed by the framework polyhedra. There are three crystallographically distinct framework sites: $6c$, $16i$, and $24k$ using the Wyckoff notation. The guest atoms occupy the $2a$ and $6d$ sites, corresponding to the pentagonal dodecahedra and the tetrakaidecahedra, respectively that form the clathrate framework. The general chemical formula can then be written as A_8E_{46}, where A represents a guest atom and E represents a framework atom.

The inorganic clathrates were first systematically synthesized by Cros et al. using thermal decomposition of alkali-silicides and -germanides.[7,8] A type I clathrate compound of composition "K_8Ge_{46}" was originally reported,[8] which exhibited semiconducting properties. Later, more detailed structural characterizations[9,10] of this compound revealed that it is actually better described by the chemical formula "$K_8Ge_{44}\square_2$," where $\square$ denotes a vacancy on the framework, specifically at the $6c$ site. As discussed in the pages that follow, this was in better accord with the measured electrical properties of this material. In this paper, we report further on the preparation and physical properties of $K_8Ge_{44}\square_2$. For the first time we present thermal conductivity data for

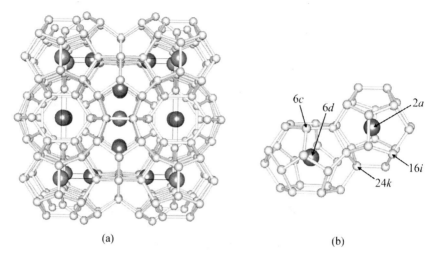

Figure 1. (a) Crystal structure of the type I clathrate. (b) The two polyhedra that form the type I clathrate framework. The crystallographic sites are labeled.

$K_8Ge_{44}\square_2$, and also find that the electrical properties of our specimen differ significantly from those reported in the literature, indicating a dependence of properties upon the synthesis conditions.

EXPERIMENTAL DETAILS

Synthesis and Sample Preparation

The $K_8Ge_{44}\square_2$ specimen in the present study was synthesized by thermal decomposition of the precursor KGe, using techniques similar to those previously reported.[8,10] First, high purity K and Ge where reacted at 650°C to form the precursor compound KGe. KGe is very reactive with air and moisture, thus all handling was performed in a nitrogen-filled glovebox. A portion of the KGe product was subsequently ground to fine powder, placed in a glass ampoule coupled to a high vacuum apparatus, and heated under vacuum (10^{-5} torr) at 440°C for 24 hours. The product consisted of very fine, grayish polycrystalline powder, which was washed with water and ethanol to remove any unreacted KGe, and then dried at 70°C under vacuum. The powder was ground further and then compacted under flowing nitrogen at 380°C and 185 MPa using a standard hot-pressing technique, resulting in a pellet approximately 83% of the theoretical X-ray density. A parallelepiped specimen for temperature dependent transport measurements was cut from the pellet using a wire saw.

Structural and Chemical Characterization

The specimen was characterized structurally employing powder X-ray diffraction, using powder ground from the hot-pressed pellet. Microstructure and chemical composition were analyzed using a scanning electron microscope (SEM), along with energy dispersive spectroscopy (EDS). Differential thermal analysis was carried out under flowing nitrogen in the temperature range 200 to 1050°C.

Transport Properties Measurements

The transport properties measurements were performed from 12 to 330 K in high vacuum using a custom designed system utilizing a closed cycle helium cryostat. All three transport coefficients (electrical resistivity, Seebeck coefficient, and thermal conductivity) were obtained on the same specimen. Electrical leads for resistivity and Seebeck coefficient measurements were soldered to nickel-plated contacts. Electrical resistivity was measured using a standard four-probe configuration. The Seebeck coefficient was measured using a transient method, in which the voltage is measured while sweeping the temperature gradient. Thermal conductivity was measured using a steady-state method.

STRUCTURAL PROPERTIES AND COMPOSITION

Fig. 2 shows a powder X-ray diffraction pattern obtained on the $K_8Ge_{44}\square_2$ specimen, post hot-pressing. The pattern is indicative of the type I clathrate crystal structure and shows the specimen to be essentially phase pure, with the exception of a very weak reflection between 27 and 28° 2θ, corresponding to a trace amount of elemental Ge that appeared after hot-pressing. The lattice parameter of the specimen was measured to be 10.667(5) Å using an internal silicon standard, in agreement with the value of 10.66771(1) Å previously reported.[10] To analyze the composition of the specimen, 11 individual grains were analyzed in a piece cut from the hot-pressed pellet using energy dispersive spectroscopy, yielding an average composition of $K_8Ge_{43.1(0.8)}$, in reasonable agreement with the nominal composition $K_8Ge_{44}\square_2$ and indicating vacancies are indeed present on the Ge framework. In calculating the chemical formula, we have

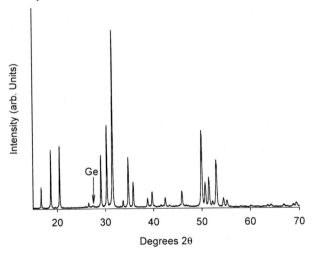

Figure 2. Powder X-ray diffraction pattern for $K_8Ge_{44}\square_2$.

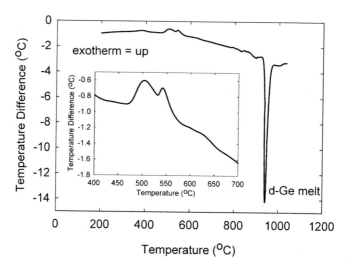

Figure 3. Differential thermal analysis (DTA) for K_8Ge_{44-2}. An enlargement of the curve in the range 400 to 700°C is shown in the inset.

assumed full occupation of the guest (K) sites, since this is empirically the case for the overwhelming majority of type I clathrates. We also note the accuracy of the chemical formula may be affected by the presence of ~ 1 wt% oxygen in the specimen, as measured by EDS. Hereafter we shall refer to the specimen by its nominal composition, $K_8Ge_{44}\square_2$.

Results from differential thermal analysis (DTA) measurements on the $K_8Ge_{44}\square_2$ specimen are shown in Fig. 3. The large endothermic transition at high temperature is the melting of elemental diamond structure Ge (d-Ge), which is present after the complete decomposition of the specimen. The curve shows at least two exothermic peaks between 400 and 600°C (inset), the first of which occurs just above 480°C. This indicates that K_8Ge_{44-2} is a meta-stable phase. The presence of multiple exothermic peaks in the curve suggests phase transitions to other K-Ge phases. von Schnering et al.[11] have studied the thermal decomposition of KGe under dynamic vacuum, and found evidence for several other binary K-Ge phases, in addition to $K_8Ge_{44}\square_2$. Several of these proposed compounds have not yet been further characterized.

TRANSPORT PROPERTIES

The temperature dependent electrical resistivity and Seebeck coefficient from 12 to 330 K for our $K_8Ge_{44}\square_2$ specimen are shown in Fig. 4. Over the majority of the temperature range studied, the resistivity monotonically increases slowly with temperature, indicating metallic-like conduction. The relatively high room temperature value of ~ 30.5 mOhm-cm however indicates the material is a poor metal, and may be a result of scattering from the vacancies on the Ge framework. The Seebeck coefficient is negative at all temperatures studied, indicating electrons are the majority carries, and obtains a value of − 80 μV/K just above room

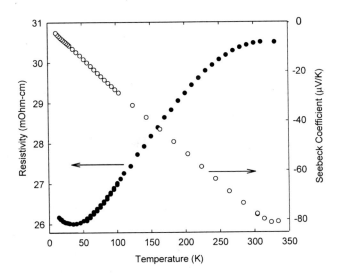

Figure 4. Temperature dependent electrical resistivity (●, left scale) and Seebeck coefficient (○, right scale) for K$_8$Ge$_{44}$□$_2$.

temperature. The apparent change in slope of both resistivity and Seebeck coefficient just above room temperature suggests dual-conduction, possibly with holes and electrons contributing simultaneously to the electrical properties.

The electrical resistivity data for our specimen differ significantly from that reported previously for K$_8$Ge$_{44}$□$_2$.[8,10] Ramachandran et al.[10] measured the resistivity on cold-pressed pellets of K$_8$Ge$_{44}$□$_2$, and found clearly activated temperature dependence. This behavior was in agreement with the temperature dependence originally reported for "K$_8$Ge$_{46}$" by Cros et al.[8] In addition, the room temperature resistivity for our specimen (~ 30 mOhm-cm) is more than a factor of 30 lower than the value (~ 1,000 mOhm-cm) reported in Ref. 10. In the following paragraphs we discuss possible reasons for these differences.

We first note that in previous work[10] resistivity measurements were carried out on cold-pressed powders. As such, it is quite difficult to ensure adequate density of the measured samples, as well as good electrical contact between the polycrystalline grains. Thermally activated transport across grain boundaries could result in perceived semiconducting behavior.[12] Adequate contact between the grains can be ensured by hot-pressing, as we have done in the present work.

The electronic properties of inorganic clathrates are often analyzed in terms of a simple rigid-band model, in which the guest atoms donate partially or completely their valence electrons to the host framework. In a material such as the hypothetical Ge$_{46}$, where each Ge participates in pseudo-tetrahedral bonding, the framework valence band should be completely occupied and the conduction band completely empty, resulting in an intrinsic semiconductor. The introduction of guest atoms into the framework would contribute electrons into the framework conduction bands, assuming the guests are partially or completely ionized. Within this model, inorganic clathrates can be considered as closed-shell Zintl phases.[13-15]

A similar model has been applied to K_8Ge_{44-2},[10] as well as other vacancy-bearing Sn clathrates.[15,16] Here, the two framework vacancies per unit cell result in a total of eight framework atoms per unit cell (the framework atoms that would be bonded to the atoms at the vacancy site) that are only three-bonded. The eight electrons donated from the framework would then be localized in non-bonding orbitals at these eight framework sites, allowing these atoms to retain their closed shells. Thus the material would again be an "intrinsic" semiconductor. This model qualitatively explained the results of Ramachandran et al., who found their $K_8Ge_{44}\square_2$ sample to be semiconducting. However, this strict picture does not explain the results shown in Fig. 3. Electrical transport measurements on type I Sn clathrates[17] revealed impurity band conduction occurs in these materials. The electrical properties of $Cs_8Sn_{44}\square_2$, a material structurally similar to $K_8Ge_{44}\square_2$, also exhibit some features (such as metallic-like temperature dependence of the electrical resistivity) similar to K_8Ge_{44-2}.[17]

Zhao et al.[18] have performed first-principles calculations for the hypothetical Ge_{46} and K_8Ge_{46} clathrates. They found that Ge_{46} behaves as an intrinsic semiconductor, while K_8Ge_{46} is expected to be metallic, as expected from the simple picture discussed above. Moreover, their calculations of the electron density showed no charge density at the K sites, indicating complete charge transfer to the framework bands.[18] We conjecture that it is possible that the stoichiometry of this material may deviate slightly from the exact K_8Ge_{44-2}, and slightly fewer vacancies on average could result in a departure from the ideal semiconducting behavior within the picture given above. A similar effect has been seen in type I Ga-Ge clathrates, in which varying the Ga to Ge atomic ratio has a significant effect upon the electrical transport.[2] We note that our $K_8Ge_{44}\square_2$ specimen was synthesized at 440°C, whereas the K_8Ge_{44-2} sample of Ramachandran et al. was synthesized at a temperature between 350 and 380°C.[10] This suggests that the synthesis conditions may have a significant effect on the properties of this material, in particular the electrical properties. A more detailed structural study should shed light on the possibility of variability of the concentration of framework vacancies.

The thermal conductivity (κ) of the $K_8Ge_{44}\square_2$ specimen is presented in Fig. 5, along with data for single crystal diamond structure Ge[19] and the type I clathrate $Sr_8Ga_{16}Ge_{30}$[2] for comparison. We find the thermal conductivity of $K_8Ge_{44}\square_2$ to be quite low for a crystalline solid, roughly 40 times lower than for single crystal germanium at room temperature and comparable in magnitude to amorphous SiO_2. As seen in Fig. 5, the room temperature value of κ is also comparable to that of another type I clathrate $Sr_8Ga_{16}Ge_{30}$; however, the temperature dependence of κ for K_8Ge_{44-2} is very different from $Sr_8Ga_{16}Ge_{30}$. $\kappa(T)$ for K_8Ge_{44-2}, showing an unambiguous low temperature peak, displays characteristics of a crystalline compound, and the temperature dependence and magnitude is similar as was found for several type I Sn clathrates.[17] In contrast, $\kappa(T)$ for $Sr_8Ga_{16}Ge_{30}$ is more akin to that for an amorphous solid. Furthermore, the clear "dip" in $\kappa(T)$ for $Sr_8Ga_{16}Ge_{30}$ was well explained by resonant scattering of the heat carrying acoustic phonons via interaction with localized guest vibration modes.[2,20]

Dong et al.[21] have previously investigated the thermal conductivity of type I Ge clathrates using molecular dynamics. From their calculations, they found an order of magnitude reduction in $\kappa(T = 300K)$ for the hypothetical Ge_{46} as compared to diamond structured Ge, and attributed this to the scaling of κ with the number of atoms in the unit cell. We expect that the low κ for $K_8Ge_{44}\square_2$ is in part due to this effect solely attributed to the framework, but also due to defect scattering from the framework vacancies, as well as phonon interactions with localized

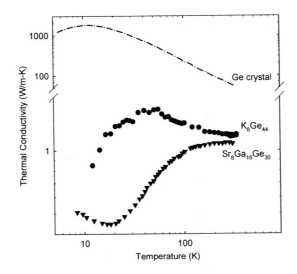

Figure 5. Thermal conductivity of K$_8$Ge$_{44}$□$_2$ (●), along with data for Sr$_8$Ga$_{16}$Ge$_{30}$ (▼) and single crystal diamond structure Ge (dashed-dotted curve). The data for Ge crystal and Sr$_8$Ga$_{16}$Ge$_{30}$ are taken from Ref. 19 and Ref. 2, respectively.

potassium guest vibrations. Dong et al.[22] have also studied the vibrational properties of Ge clathrates with various alkali and alkaline-earth guests using density functional theory. Their results indicate a weaker interaction between guest and framework vibrations for alkali guests as compared to alkaline earth guests, resulting in less pronounced phonon scattering by the guest atom vibrations. This is in agreement with the results of Figure 5, which show a strong dip in $\kappa(T)$ for Sr$_8$Ga$_{16}$Ge$_{30}$, whereas $\kappa(T)$ for K$_8$Ge$_{44}$□$_2$ has a character more indicative of a crystalline solid. We therefore conclude that the low thermal conductivity for K$_8$Ge$_{44}$□$_2$ relative to diamond structure Ge is largely due to a combination of the enlarged clathrate unit cell, as well as phonon scattering by framework vacancies and localized potassium guest vibrations, although the latter is less efficient as compared to a material such as Sr$_8$Ga$_{16}$Ge$_{30}$.

CONCLUSIONS

Clathrate compounds continue to be of interest as potential thermoelectric materials. As part of a continuing exploration of these materials and their properties, we have synthesized and characterized K$_8$Ge$_{44}$□$_2$, a type I Ge clathrate with K inside the framework polyhedra. The thermal conductivity of this compound was found to be very low, and comparable in magnitude at room temperature to the thermal conductivities of other clathrate materials. Our K$_8$Ge$_{44}$□$_2$ specimen was found to behave as a poor metal, in contrast to results from the literature which found K$_8$Ge$_{44}$□$_2$ to be semiconducting. We suggest that the precise stoichiometry of this compound may vary depending upon the synthesis conditions, and this could be the cause of variation in electrical properties. A further detailed structural and synthetic study will help elucidate the chemical and physical properties of this material and their dependence upon conditions of preparation.

ACKNOWLEDGEMENTS

This work was supported in part by the Department of Energy under Grant Number DE-FG02-04ER46145. The authors gratefully acknowledge D. Guarrera from JEOL, as well as Oxford EDS, for SEM and EDS measurements. MB is thankful for continuing support from the University of South Florida Presidential Doctoral Fellowship.

REFERENCES

[1] P. Rogl, in: D.M. Rowe (Ed.), *Thermoelectrics Handbook: Macro to Nano*, CRC Press, Boca Raton, 2006, p. 32-1.

[2] G.S. Nolas, J.L. Cohn, G.A. Slack, and S.B. Schujman, *Appl. Phys. Lett.* **73**, 178 (1998).

[3] G.S. Nolas, in: D.M. Rowe (Ed.) *Thermoelectrics Handbook: Macro to Nano*, CRC Press, Boca Raton, 2006, p. 33-1.

[4] H. Kawaji, H. Horie, S. Yamanaka, and M. Ishikawa, *Phys. Rev. Lett.* **74**, 1427 (1995).

[5] S. Paschen, W. Carrillo-Cabrera, A. Bentien, V.H. Tran, M. Baenitz, Yu. Grin, and F. Steglich, *Phys. Rev. B* **64**, 214404 (2001).

[6] G.T. Woods, J. Martin, M. Beekman, R.P. Hermann, F. Grandjean, V. Keppens, O. Leupold, G.J. Long, and G.S. Nolas, *Phys. Rev. B* **73**, 174403 (2006).

[7] J.S. Kasper, P. Hagenmuller, M. Pouchard, and C. Cros, *Science* **150**, 1713 (1965).

[8] C. Cros, M. Pouchard, and P. Hagenmuller, *J. Solid State Chem.* **2**, 570 (1970).

[9] Llanos, J., Doctoral Dissertation, University of Stuttgart, 1983.

[10] G.K. Ramachandran, P.F. McMillan, J. Dong, and O.F. Sankey, *J. Solid State Chem.* **154**, 626 (2000).

[11] H.G. von Schnering, M. Baitinger, U. Bolle, W. Carrillo-Cabrera, J. Curda, Y. Grin, F. Heinemann, J. Llanos, K. Peters, A. Schmeding, and M. Somer, *Z. anorg. allg. Chem.* **623**, 1037, (1997).

[12] N.F. Mott and E.A. Davis, *Electronic Processes in Non-Crystalline Materials*, Oxford University Press, New York, 1971.

[13] E. Zintl, *Angew. Chem.*, **52**, 1 (1939).

[14] W. Klemm and E. Busmann, *Z. Anorg. Allg. Chem.* **319**, 297 (1963).

[15] S.M. Kauzlarich (Ed.), *Chemistry, Structure and Bonding of Zintl Phases and Ions*, VCH, New York, 1996.

[16] J.-T. Zhao and J.D. Corbett, *Inorg. Chem.* **33**, 5721 (1994).

[17] G.S. Nolas, J.L. Cohn, J.S. Dyck, C. Uher, and J. Yang, *Phys. Rev. B* **65**, 165210 (2002).

[18] J. Zhao, A. Buldum, J.P. Lu, and C.Y. Fong, *Phys. Rev. B* **60**, 14177 (1999).

[19] C.J. Glassbrenner and G.A. Slack, *Phys. Rev.* **134**, A1058 (1964).

[20] J.L. Cohn, G.S. Nolas, V. Fessatidis, T.H. Metcalf, and G.A. Slack, *Phys. Rev. Lett.* **82**, 779 (1999).

[21] J.J. Dong, O.F. Sankey, and C.W. Myles, *Phys. Rev. Lett.* **86**, 2361 (2001).

[22] J.J. Dong, O.F. Sankey, G.K. Ramachandran, and P.F. McMillan, *J. Appl. Phys.* **87**, 7726 (2000).

Transparent Electronic Ceramics

ADVANCED INDIUM TIN OXIDE CERAMIC SPUTTERING TARGETS AND TRANSPARENT CONDUCTIVE THIN FILMS

Eugene Medvedovski, Neil A. Alvarez, Christopher J. Szepesi, Olga Yankov
Umicore Indium Products
50 Sims Ave., Providence, RI 02909, USA

Maryam K. Olsson
Umicore Maerials AG
Altelandstrasse 8, FL-9496, Balzers, Liechtenstein

ABSTRACT

Highly electrically conductive and transparent thin films are widely used as electrode layers in optoelectronic devices, such as flat panel displays, solar cells, electrochromic devices and antistatic conductive films. The films may be produced by magnetron sputtering technique that requires a fine-tune deposition process and high quality sputtering targets. Dense indium tin oxide (ITO) ceramics are one of the most reliable materials for sputtering targets. Umicore Indium Products (UIP) has a strong experience in the development and manufacturing of ITO ceramics using in-house prepared In_2O_3 powders. The challenges of ceramic composition and manufacturing are considered; they include the use of high quality starting materials, especially In_2O_3 powders, with respect to purity, morphology and sinterability, manufacturing routes and sintering process. The obtained ITO ceramics have a high purity, uniform microcrystalline structure, high density up to 99+% of TD and low specific electrical resistivity. Using manufactured sputtering targets, which may be made as large tiles with areas of 1000-1700 cm^2, and a fine-tune optimized sputtering process, high quality electrically conductive transparent films with nano-crystalline or amorphous structures are obtained. Physical properties of ITO ceramics and thin films are reported.

INTRODUCTION

Highly electrically conductive and transparent thin films are widely used as electrode layers in optoelectronic devices, such as in flat panel displays (FPD), e.g. liquid crystal displays (LCD), organic light-emitting diodes (OLED) and some others, touch panels, solar cells, electrochromic devices, as well as antistatic conductive films and low-emission coatings [1-5]. The films are commonly produced by conventional DC magnetron sputtering on glass or polymer substrates, requiring a fine-tuned deposition process and high quality sputtering targets. One of the most reliable and suitable materials, among different transparent conductive oxides (TCO), for sputtering targets is indium-tin oxide (ITO) ceramics. These ceramics are formed by the doping of some amounts of tin oxide to indium oxide, that results in modification (distortion) of crystalline lattice of indium oxide and in enhance its electrical conductivity. The ceramics should be of high purity with a uniform microcrystalline structure. They should possess high density (99+% of TD) to maximize the useful life of the targets and high electrical conductivity, and their use in sputtering system should provide a formation of crystalline or amorphous electrically conductive transparent films without structural defects. Due to the present need in large area optoelectronic devices with high quality films, the ceramic targets are desirable to be as large as possible, i.e. dense monolithic tiles with areas up to 1500-1700 cm^2, which are used for assembling large-sized sputtering targets, are required, especially for FPD industry. The manufacturing of these large-sized fully dense products is quite challenging for commercial ceramic processing.

Required electrical conductivity and transparency of ITO films are defined by compositional and structural features of the ceramics and by sputtering process parameters. Regarding ceramics, the structure of the crystalline lattice of In_2O_3 modified by the dopant and oxygen deficiency in the lattice have a high importance, i.e. the content of SnO_2, substitution of In^{3+} by Sn^{4+} in the cation sites (that results in the donation of free electrons to the lattice and provides n-type conductivity) and densification of ceramics are among crucial factors [6-8]. Due to the lattice defects caused by interstitial atoms or vacancies and oxygen deficiency, theoretical density of ITO ceramics is not precisely defined; e.g. it is considered as 7.13-7.16 g/cm^3 for the ITO 90/10 ceramic compositions.

The quality of DC magnetron sputtered thin films is generally superior when the ceramic targets have higher density; also higher density targets promote deposition rate [4]. Dense ceramic targets have higher resistance against sputtering erosion and nodule formation. In particular, nodules ("black spots"), which may be considered as indium sub-oxide, occur during sputtering on the periphery of erosion race track of targets and tend to cause electrical arcing; they deteriorate properties of the films and should be periodically removed during processing. The nature of the nodules formation is complex, and its mechanism has not yet been completely understood [4, 9, 10]. However, based on the experimental results, B.L. Gehman et al [4] noted that very high ceramic target purity is not the major condition to attain the films with lower resistivity, i.e. ultrahigh purity grade targets would be only a small advantage of the film quality. Sputtering parameters may have a greater influence on the film quality [10].

One of the most widely used ITO compositions is 90/10 ITO, i.e. with an approximate wt.-% ratio of 90/10 between In_2O_3 and SnO_2; this ceramic composition provides high quality conductive and transparent films required for the optoelectronic applications. However, some other ITO compositions, such as 98/2, 97/3, 95/5, 80/20, are also used in optoelectronic applications.

The development and study of ITO and some other In_2O_3-based ceramics are under ongoing attention of the ceramic manufacturers and the TCO users [1-10]. Despite the numerous studies, it is not enough data for the ceramics produced on a commercial basis. In the present work, the challenges of the ceramic composition and manufacturing of ITO sputtering targets are considered; they include the use of high quality starting materials, especially In_2O_3 powders with respect to purity, morphology and sinterability, manufacturing routes and sintering process. ITO ceramic tiles with areas up to 1200-1700 cm^2 (with a variety of dimensions) and densities of 99+% of TD are currently produced by Umicore Indium Products (UIP) using in-house prepared In_2O_3 powders. Properties of the ITO ceramics manufactured in the industrial conditions and the films obtained from these ceramics deposited by DC magnetron sputtering are reported.

EXPERIMENTAL
Starting Materials and Processing

High-purity commercially produced In_2O_3 and SnO_2 powders are used as the main starting materials for production of ITO ceramics. The In_2O_3 powders are manufactured by UIP using a proprietary process from pure indium via its acidation with subsequent neutralization and precipitation of $In(OH)_3$. Then the prepared $In(OH)_3$ is calcined at a proper temperature. Usually hydrochloric acid is used for acidation, and the In_2O_3 powders prepared via this route are denoted as type II. Each lot of starting In and prepared $In(OH)_3$ and In_2O_3 powders are qualified by chemical analyses and powder characterization. Properties of resultant In_2O_3 powders are summarized in Table 1, and the typical particle size distribution of In_2O_3 powders is displayed in Fig. 1. The In_2O_3 powders have cubic shape, and they are generally aggregated (Fig. 2). Properties of SnO_2 powders obtained from external suppliers are also summarized in Table 1.

Table 1. Properties of Starting Powders for ITO Ceramics

Material	Purity, %	Particle size distribution, μm			Specific surface (BET), m^2/g
		d10	d50	d90	
$In(OH)_3$	99.99	1-3	7-12	15-20	8-12
In_2O_3, type ll	99.99	0.8-0.9	3-5	7-8	0.7-1.0
SnO_2	99.9	0.1-0.2	0.3-0.9	2-3	4-9
ITO slip	-	0.1-0.2	0.4-0.6	1.5-2.5	5-9

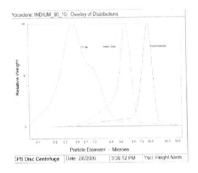 

Fig. 1. Particle Size Distribution for $In(OH)_3$ and In_2O_3 Powders and ITO Slip

Fig. 2. SEM Image of In_2O_3 Powder

Starting materials are mixed and milled using a specially developed wet process to obtained slurries with relatively high specific gravities and workable viscosities (the slurry compositions with selected dispersing and binding agents have been developed). The shaped ceramic bodies are dried, cut in a "green" state and fired in the high-temperature electric furnaces using specially designed loading and firing conditions. The optimized firing cycle (firing temperature is below $1600°C$) and firing conditions provide practically full densification (up to 99.5% of TD). Then fired tiles are cut and ground with diamond tooling in order to provide precise dimensions, flatness and surface quality, which are required for the back face metallization, bonding and magnetron sputtering processes. For example, roughness of the ceramics Ra is attained to be not greater than 0.45 μm. It should be noted that a multi-step process control during powder preparation and ceramic manufacturing is maintained that provides a high purity of ceramic targets.

Sputtering
ITO films were deposited from ITO ceramic targets with dimensions up to 381x127x6 mm cut from actually producing tiles onto glass substrates Corning No. 1737F using an industrial vertical planar DC magnetron sputtering system (LLS EVO II). The base pressure in

the process module was below 5×10^{-5} Pa. Typically a power density of 3.1 W/cm^2 was applied during the sputtering runs. The sputtering pressure was in the range of 0.3-0.6 Pa that could be adjusted using a mass flow controller. The sputtering process was run in Ar atmosphere with an addition of 0-5% of O$_2$ reactive gas. The sputtering runs were conducted at ambient and elevated temperatures (up to 200°C); IR-radiation lamps installed in the process module were applied to maintain uniform substrate heating. Post-deposition heat treatment (annealing) of the coatings produced at ambient temperature was carried out using an electric furnace in air atmosphere. Annealing was carried out at temperature of 200°C during 60 min. The parameters of the sputtering process were optimized in order to obtain high-quality films.

Characterization

Powders were characterized for particle size distribution and specific surface using the Sedigraph and CPS Disc Centrifuge instruments and the Brunauer-Emmett-Teller (BET) method, respectively. The morphology of the powders was studied by high resolution scanning electron microscope (SEM). Phase composition and microstructure of the fired ceramics and sputtered films were studied using glancing incidence angle X-ray diffraction (XRD) and SEM and transmission electron microscopy (TEM) under different magnifications using "as-received" or etched samples. Thermal gravimetric analysis (TGA) was conducted in the range of 20-1400°C using a standard procedure. Oxygen content in the ITO ceramics was determined using TGA. This method is based on the calculation of the weight decrease after reduction of the pulverized sintered ITO ceramic powder since it is assumed that the weight decrease is equivalent to the oxygen content in ITO when In and Sn are fully reduced.

Density of ceramic tiles was measured by the water immersion method based on Archimedes law. Young's modulus was tested using the resonant frequency method in accordance with ASTM C885. Sonic velocity was determined based on the formula connected Young's modulus and sonic velocity: $E = c^2 d(1+p)(1-2p)/(1-p)$, where E is Young's modulus, c is sonic velocity, d is density, p is Poisson ratio that is also determined by the ASTM 885. Flexural strength (four-point loading) was tested in accordance with ASTM C1161. Thermal diffusivity α was measured by the laser flash technique in accordance with ASTM E1461. Specific heat (heat capacity) C_p was tested using differential scanning calorimeter (ASTM E1269). Thermal conductivity λ values were calculated using a formula: $\lambda = \alpha C_p d$ (where d is density). Thermal properties were determined in the temperature range of 20-250°C. Coefficient of thermal expansion (CTE) was determined in the temperature range of 20-1000°C using a quartz dilatometer. Specific electrical resistivity was determined using the four-point probe measuring unit. The test ceramic samples were cut from actual tiles-targets for the dimensions required by the appropriate testing procedures.

The ITO film thickness was optically determined using a J.Y. Horiba reflectometer analyzer. Specific electrical resistivity of the films also was determined using a Jandel four-point probe measuring unit. The transmittance in the visible range from 400 to 800 nm wavelength was measured using a Perkin Elmer optical analyzer. Film stress was determined for the selected films deposited onto Si wafers with 100 mm diameter using a standard procedure for flat panel displays using cantilever technique. The samples with deposited layers were subjected to a bending stress, and the bending contour was determined using an α-step profilometer.

RESULTS AND DISCUSSION

The weight ratio between In_2O_3 and SnO_2 is selected in accordance with requirements of electrical properties of the ceramics and electrical properties and transmittance of the films. Based on numerous studies, the ratio of 90/10 generally provides the lowest electrical resistivity and high transmittance of the films prepared by sputtering techniques; however some other compositions may also be used for optoelectronic industrial applications.

Slip composition of ITO ceramics is based on the optimal ratio between solid and liquid phases; the latter consists of water and the specially selected dispersants providing low viscosity with a relatively high specific gravity and a proper pH level. Based on the conducted studies focused on the selection of the dispersant and binder system with a proper pH level, the ITO slips have approximate solid contents of 80-84 wt.-% (depending on the slip compositions).

In order to achieve a high level of homogenization of the ITO mix components and the slips and sinterability of the ceramics, a wet mechanochemical activation process is used. An extremely low level of impurities, which may be introduced into ceramic mix during milling process (e.g. due to wear of grinding media and lining of the milling equipment) has to be maintained. In order to achieve that, the grinding media and the lining should provide the highest wear resistance as possible, and the impurities incoming with the milling process should have a smaller effect on deterioration of sinterability of ITO and physical properties of sintered ceramics and deposited thin films. The optimized milling process provides particle size distribution and specific surface of the ITO slips, which are suitable for different types of ceramic processing, even for the bodies with quite large dimensions, and to obtain adequate sinterability of the ceramics.

Densification of ITO ceramics depends on properties and morphology of starting materials, composition, e.g. a weight ratio of In_2O_3/SnO_2 in the mix, ceramic processing route and firing conditions. The ITO ceramic formation occurs via solid state sintering. SnO_2 may segregate preferentially around the voids in the interior of the sintered body, so it is desirable for better densification if SnO_2 particles are smaller as possible and if SnO_2 is well distributed between In_2O_3. This may be achieved by optimization of the mixing-milling process. The morphology of the starting powders affects the sinterability of ITO ceramics. It was found that if some amounts of In_2O_3 powder do not have a cubic shape due to some imperfectness of the In_2O_3 preparation, the ceramics have lower density. This is in a good correlation with the study conducted by B.-C. Kim et al [11], who found that the transformation of rhombohedral In_2O_3 to cubic induces coarsening of grains and formation of voids in microstructure retarding densification of the ceramics. In_2O_3-based ceramic compositions generally have a low sinterability. It is related to the partial dissociation and vaporization of In_2O_3 and SnO_2 at elevated temperatures in accordance with reactions [12-14]:

In_2O_3 (s) $\rightarrow$ In_2O (g) + O_2 (g) ($<1300^oC$)
In_2O_3 (s) $\rightarrow$ $2InO$ (g) + $1/2O_2$ (g) ($>1300^oC$)
SnO_2 (s) $\rightarrow$ SnO (g) + $1/2O_2$ (g)

Partial vaporization of the oxides at elevated temperatures is confirmed by the results of TGA if the analysis is conducting at the lower oxygen environment (Fig. 3). Due to this fact, a temperature increase might not assist full densification if some conditions are not maintained. In order to achieve high densification (98+% of TD), hot pressing or an addition of sintering aids to the compositions or applying some other specific technological methods have to be used in order to modify firing process and to improve sinterability of ITO ceramics. Based on the conducted studies, a specially designed firing profile and oxygen-assisting firing conditions are used for sintering of the ITO ceramics. Density of ITO ceramic tiles fired in industrial conditions achieved 7.12 g/cm^3 (90/10 composition), i.e. up to 99.5% of TD (7.14 g/cm^3), with a good reproducibility. The density values strongly depend on the size of the tiles and the firing conditions. As the smaller the tiles dimensions, higher density of the tiles may be easier obtained. However, these high density values can be achieved not only for the tiles with areas of 500-900 cm^2, but also for the tiles with areas of up to 1200-1700 cm^2. Fired large tiles with the above mentioned values of density produced at UIP are demonstrated in Fig. 4. Compared to hot pressing, the using firing conditions have valuable advantages with respect to lower cost of manufacturing equipment and lower running cost and greater capacities of the furnaces for mass production, including a possibility to fire large-sized products.

Fig. 3. TGA curve of ITO (90/10) Ceramics

Fig. 4. ITO (90/10) Sputtering Target with a Total Length of 1400 mm

The ITO ceramics studied in this work have the major crystalline phase of In$_2$O$_3$ (bixbyite). However, XRD analysis indicates the presence of a secondary phase In$_4$Sn$_3$O$_{12}$ in ITO 90/10 and 80/20 ceramics that correlates with the literature data with respect to the presence of this phase in the ITO ceramics with a content of 6 at.% or more of Sn [8, 15, 16]. In many cases, the detection of this secondary phase is rather hard since its peaks on the difractograms are overlapping with the peaks of the major In$_2$O$_3$ phase. The formation of this secondary phase may be considered as positive from the ITO ceramic densification standpoint because this new phase forming during sintering occupies the "space" between In$_2$O$_3$ grains. Due to the formation of the In$_4$Sn$_3$O$_{12}$ phase, it is much easier to obtain density values of 7.10-7.14 g/cm^3 for the 80/20 composition, although TD of this composition is lower. Crystallization of SnO$_2$ phase was not detected that is in agreement with other studies [8, 13, 15]. Microstructure of the ITO ceramics made by UIP is dense, rather uniform, and it consists of the grains with sizes from 2 to 20 μm mostly with a cubic shape (Fig. 5), which consist of the crystallites with sizes of 25-45 nm. The sizes of crystallites were determined via XRD analysis using the Scherrer formula for calculation. Probably, a formation of the secondary phase promotes inhibition of the In$_2$O$_3$ grain growth that may be confirmed by the comparison of the grain sizes of the studied materials. As can be seen from the SEM images, the grain sizes of ITO 95/5 ceramics are 10-20 μm (Fig. 5a), which does not contain the secondary phase, are larger than 90/10 (Fig. 5b) and 80/20 (Fig. 5c) ceramics (grain sizes are 5-10 μm and 2-7 μm, respectively). The grains cleavage can be seen at the SEM images of the fracture of the samples that may be an evidence of a high densification of the ceramics since the fracture occurs through the grains. Irregular small intergranular pores are uniformly distributed, but they are not interconnected.

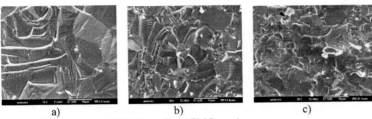

a)   b)   c)

Fig. 5. Microstructure of ITO Ceramics (SEM Image)
a) 95/5; b) 90/10; c) 80/20

Properties of the ITO ceramics produced by UIP are performed in Table 2. Mechanical strength of these ceramics is on the moderate level; however, ITO ceramics are not intended for structural applications. Young's modulus and sonic velocity, additionally to structural properties, indicate a "level" of densification, especially on the presence and quantity of closed pores and macrodefects, and their data may be useful for the comparison of the ceramics with the same composition, e.g. ITO materials. However, the determination of sonic velocity using the above formula may not be enough accurate for ITO ceramics because of relatively wide Poisson ratio variations for these materials.

The values of specific electrical resistivity of ITO ceramics are generally lower for the compositions of 90/10 comparatively with some other ITO compositions, e.g. they are ranging of $(1.3-1.7) \times 10^{-4}$ Ohm.cm for the compositions of 90/10 vs. $(2.5-2.7) \times 10^{-4}$ and $(1.6-1.8) \times 10^{-4}$ Ohm.cm for the compositions of 80/20 and 95/5, respectively. However, the measure of electrical resistivity is not a 100% indicator of the ITO composition, especially if SnO_2 content is in the range of 5-12%. Densification of ITO ceramics has higher influence on the electrical resistivity (the ceramics of the same composition but with higher density usually demonstrate lower values of specific electrical resistivity) than their ultrahigh purity. However, total high contents of impurities (such as more than 1000 ppm) may result in not only the undesirable distortion of the In_2O_3 crystalline lattice, but also in the formation of the electrically insulating layers between In_2O_3 grains, i.e. may enhance the ceramic electrical resistivity and, consequently, enhance the film resistivity. In this case, the consideration of impurities and their influence on electrical properties of the ITO ceramics should be "selective", e.g. the formation of alkali silicate or alkali earth silicate or some other glassy phases should be eliminated; at the same time, some other impurities do not affect electrical properties of ITO ceramics.

Table 2. Some Physical Properties of Studied UIP ITO Ceramics

Property	90/10	95/5	80/20
Density, g/cm^3	7.07-7.12	7.07-7.12	7.07-7.14
Oxygen content, %	17.55-17.90	17.40-17.46	17.90-17.97
Flexural strength, MPa	150-180	-	-
Young's modulus, GPa	160-190	-	-
Poisson ratio	0.285-0.335	-	-
Sonic velocity, km/s	5500-6400	-	-
Specific electrical resistivity, Ohm.cm	$(1.3-1.6) \times 10^{-4}$	$(1.6-1.8) \times 10^{-4}$	$(2.5-2.7) \times 10^{-4}$
$CTE \times 10^6$, 1/K 20-200°C 20-700°C	7.0-7.5 8.3-8.6	7.0-7.3 8.2-8.6	7.4-7.6 8.5-8.6
Thermal diffusivity, cm^2/s 20°C 250°C	0.042-0.046 0.034-0.038	0.045-0.050 0.038-0.042	0.028-0.032 0.024-0.026
Heat capacity, W.s/g-K 20°C 250°C	0.36-0.37 0.40-0.41	0.36-0.37 0.42-0.43	0.36-0.37 0.43-0.44
Thermal conductivity, W/m-K 20°C 250°C	11-12 10-11	12.5-13 12-12.5	8-8.5 7.5-8

The values of electrical resistivity of ITO ceramics may vary in a rather wide range depending on the content of the lower conductive phase $In_4Sn_3O_{12}$. This phase, as shown above, is formed at the Sn content greater than 6%, and its content and solubility of Sn in In_2O_3 may depend on the processing features, such as mixing-milling and, especially, firing conditions (e.g. temperature distribution in the kiln, oxygen level, etc.). Practically, oxygen level and sintering process also depend on the size of the ITO ceramic tiles, on their loading in the furnace and "thermal mass" of the ceramics (related to the tile sizes and amount of tiles in the furnace), which are varied from firing to firing in the manufacturing conditions. I.e. depending on these factors, the content of the forming $In_4Sn_3O_{12}$ phase and electrical conductivity of the ceramics may vary. Elevated values of electrical resistivity of the 80/20 composition are explained by the crystallographic structure of the material and a greater extent of the lower conductive $In_4Sn_3O_{12}$ phase. However, the presence of this $In_4Sn_3O_{12}$ phase positively affects density of ITO ceramics, i.e. the influence of the phase composition and structure of ITO ceramics on their properties and quality of the film is rather complex.

It is hard to correlate the influence of microstructure (grain size and grain size distribution) on electrical properties of ITO ceramics. There is no indication in literature regarding this influence, even for the small size samples processed and fired at very uniform laboratory conditions. Considering real production conditions, due to the variations in starting powders preparation and firing conditions, the structure may vary from firing to firing, especially if large-sized tiles are fired, and this influence of microstructure may be hardly observed. Only a small tendency of a conductivity increase with a grain size growth might be noticed.

Thermal conductivity of ITO ceramics has to be as high as possible, and is related with minimizing thermal tensile stresses naturally occurring in the targets during sputtering; otherwise, related cracks occurring may be a cause of non-uniform deposition of ITO films. The values of thermal diffusivity and thermal conductivity of ITO ceramics are higher for the materials with higher contents of In_2O_3. The influence of densification on the thermal properties of the ceramics is similar to its influence on electrical conductivity; the higher the density, the greater the thermal diffusivity, the heat capacity and the thermal conductivity. The samples with higher electrical conductivity also demonstrate higher thermal conductivity. The values of thermal diffusivity and thermal conductivity decrease with a temperature increase, and the change of thermal conductivity vs. temperature has a linear character, but the change of thermal diffusivity vs. temperature does not have such behavior (Fig. 6). Both electrical and thermal conductivity tend to be higher for the ITO samples with larger grains; probably, it may be explained that if the ceramics have larger grains, less grain boundaries promote less resistance to electrical and thermal flows (with respect to electrical properties, this trend may be noted for the samples with significant difference in the grain sizes). The values of CTE of ITO ceramics with different contents of SnO_2 are on the same level that allows using similar compositions for metallization and bonding processes of ITO targets.

The thin films with an average thickness of 90-100 nm obtained by the conventional DC magnetron sputtering at different conditions were analyzed for their compositions, microstructure and physical properties. The study results for the films obtained from the targets made from 90/10 ITO ceramics as the most reliable composition are reported. A wide process "window", a stable sputtering process and high quality layers were obtained by adjusting the amount of oxygen inserted to the system and some other sputtering parameters. The glancing incidence angle XRD spectra of selected ITO layer structures were recorded and compared with JCPDS database #6-0416. The ITO films deposited at room temperature ("as-deposited") had, based on XRD analysis, an amorphous structure (a-ITO); however, small amounts of crystallites randomly distributed were also observed. A polycrystalline structure of cubic shape In_2O_3 grains is inherent for the films deposited at elevated temperature of substrate. The films produced after annealing at 200°C also had very homogeneous polycrystalline structures (p-ITO) consisted of the In_2O_3 phase. These crystalline structures had equiaxed (cubic) grains with average sizes of 15-30 nm densely packed (Fig. 7). Only occasional micropores might be observed at the triple junctions of the grains. The annealing process promotes crystallization and, subsequently, an increase of density of the films.

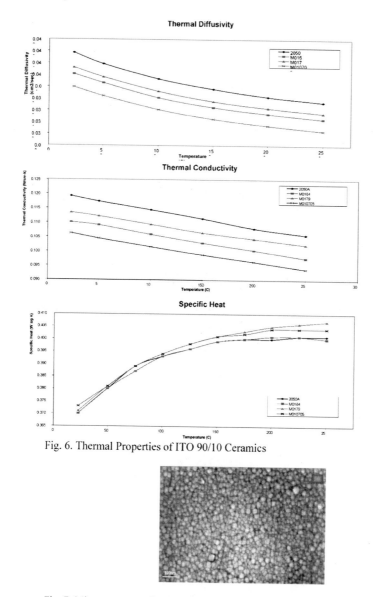

Fig. 6. Thermal Properties of ITO 90/10 Ceramics

Fig. 7. Microstructure of ITO (90/10) Film Prepared by DC Magnetron Sputtering (Deposition at Room Temperature with Subsequent Annealing at 200°C during 1 hr.)

Specific electrical resistivity of the amorphous ("as-deposited") films was as low as about 600 μOhm.cm at the oxygen flow of approximately 1.6% (Fig. 8 a). After annealing, specific electrical resistivity of the nanocrystalline ITO structures was lowered to 210 μOhm.cm that is well accepted for various FPD applications. The transparency of the layers at 550 nm wavelength was approximately 85% with insertion of approximately 1.8% of oxygen for "as-deposited" films, while it was increased to more than 90% after 1 hr annealing in air (Fig. 8 b). The results are in a good agreement with the previous work [17] when an annealing process results in a relaxation of disorder bands as well as crystallization of the amorphous network, hence increasing electrical conductivity and optical transmittance of the layer. Deposition on the heated substrate provided obtaining of the same level of specific electrical resistivity and optical transmittance as the annealed layers due to the crystallization on the hot substrate.

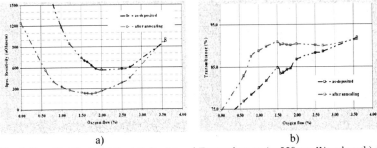

a)   b)

Fig. 8. Specific Electrical Resistivity (a) and Transmittance (at 550 nm Wavelength) (b) of ITO Films Prepared from 90/10 ITO Ceramic Target by DC Magnetron Sputtering Deposition (blue line – "as-deposited", red line – after annealing)

The selected samples were evaluated for the film stress. The "as-deposited" films are considered as being under compressive stress that occurs from the thermal expansion mismatch between ITO layer and the Si wafer substrate [17]. The lowest value of approximately -200 MPa was measured. After annealing of the amorphous ITO film, the film stress converted from compression to tensile with a value of approximately +200 MPa. The transition of compressive stress of a-ITO to tensile stress of p-ITO after annealing may be due to an enormous densification of the amorphous structure transforming to the crystalline phase. Typically, the specifications for stress in ITO coatings for use in LCD industry are set at lower than +/-500 MPa compressive/tensile [17], i.e. the obtained results are in a good agreement with industry requirements.

It was hard to define a strong correlation between density of ITO ceramics and macroscopic properties of the ITO films. Based on the studies conducted by K. Utsumi et al [5], an increase of density of ITO ceramics from 90 to 99% of TD resulted in a slight decrease of the film resistivity due to a slight increase of the carrier concentration, and this change of resistivity was noted when density of samples increased from 97 to 99% of TD; however, there was not found an influence on transmittance of the films. However, it should be noted that all samples, which were undergone for the sputtering tests in the present work, had a high level of density (99% of TD or greater) that is required by the FPD industry. This high level of density of ceramics has to be maintained in order to achieve not only high quality of film properties, but also to minimize occurrence of the defects in the films and to maximize the sputtering efficiency in industrial conditions, especially in the case of the use of large area targets.

CONCLUSION

The developed technology of ITO ceramics using in-house prepared starting In_2O_3 powders allows to manufacture on the industrial basis high-quality tiles with different dimensions and with the area up to 1700 cm^2, which are successfully used for planar sputtering targets. Fig. 4 illustrates a large target assembled from the UIP large center tiles (with a length of almost 500 mm) and smaller side tiles with an increased (15 mm) thickness. The manufactured ceramics have density up to 99.5% of TD for the products of various dimensions. Due to the selected compositions, high uniformity and densification, the obtained ceramics possess low electrical resistivity and acceptable structural and thermal properties. As a result, the DC magnetron sputtered films obtained with fine-tuned sputtering process parameters reveal low specific electrical resistivity and high transmittance (90% or greater at the 550 nm wavelength) required for flat panel display and some other optoelectronic industrial applications.

ACKNOWLEDGEMENTS

The authors are grateful to Dr. G. Huyberechts (Umicore RDI, Belgium) for the helpful discussions and assistance in structural analyses. Assistance of Fraunhofer Institute for Thin Films and Surface Technology (Germany) is appreciated greatly for conducting of XRD and high resolution SEM analysis of thin films.

REFERENCES

1. J.L. Vossen, "Transparent Conducting Films", *Phys. Thin Films*, Ed. By G. Haas, M.H. Francombe, and R.W. Hoffman (Academic, New York) **9**, 1-71 (1977)

2. D.S. Ginley, C. Bright, "Transparent Conducting Oxides", *MRS Bulletin*, **8**, 15-18 (2000)

3. I. Hamberg, C.G. Granquist, "Evaporated Sn-Doped In_2O_3 Films: Basic Optical Properties and Applications to Energy-Efficient Windows"; *J. Appl. Phys.*, **60**, R123-160 (1986)

4. B.L. Gehman, S. Jonsson, T. Rudolph, et al, "Influence of Manufacturing Process of Indium Tin Oxide Sputtering Targets on Sputtering Behavior"; *Thin Solid Films*, **220**, 333-336 (1992)

5. K. Utsumi, O. Matsunaga, T. Takahata, "Low Resistivity ITO Film Prepared Using the Ultra High Density ITO Target", *Thin Solid Films*, 334, 30-34 (1998)

6. 8. G. Frank, H. Kostlin, "Electrical Properties and Defect Model of Tin-Doped Indium Oxide Layers", *Applied Physics A*, 27, 197-206 (1982)

7. P.A. Cox, W.R. Flavell, R.G. Egdell, "Solid-State and Surface Chemistry of Sn-Doped In_2O_3 Ceramics", *J. Solid State Chem.*, **68**, 340-350 (1987)

8. J.L. Bates, C.W. Griffin, D.D. Marchant, et al., "Electrical Conductivity, Seebeck Coefficient and Structure of In_2O_3-SnO_2". *Amer. Ceram. Soc. Bull.*, **65**, N.4, 673-678 (1986)

9. A.D.G. Stuwart, M.W. Thompson, "Microphotography of Surfaces Eroded by Ion Bombardment", *J. Mater. Sci.*, **4**, 56-60 (1969)

10. S. Ishibashi, Y. Higuchi, Y. Oka, et al, "Low Resistivity Indium-Tin Oxide Transparent Conductive Films. II. Effect of Sputtering Voltage on Electrical property of Films", *J. Vac. Sci. Technol. A*, **8** (3), May/June, 1403-1406 (1990)

11. B.-C. Kim, S.-M. Kim, J.-H. Lee, et al, "Effect of Phase Transformation on the Densification of Coprecipitated Nanocrystalline Indium Tin Oxide Powders", *J. Amer. Ceram. Soc.*, **85**, N. 8, 2083-2088 (2002)

12. J.H.W. de Wit, "The High Temperature Behavior of In_2O_3", *J. Solid State Chemistry*, **13**, 192-200 (1975)

13. J.H.W. de Wit, M. Laheij, P.E. Elbers, "Grain Growth and Sintering of In_2O_3", *J. Solid State Chemistry*, **13**, 143-150 (1975)

14. R.H. Lamoreaux, D.L. Hidenbrand, L. Brewer, "High-Temperature Vaporization Behavior of Oxides II. Oxides of Be, Mg, Ca, Sr, Ba, B, Al, Ga, In, Tl, Si, Ge, Sn, Pb, Zn, Cd, and Hg", *J. Phys. Chem. Ref. Data*, **16**, N. 3, 419-443 (1987)

15. N. Nadaud, N. Lequeux, M. Nanot, et al., "Structural Studies of Tin-Doped Indium Oxide (ITO) and $In_4Sn_3O_{12}$", *J. of Solid State Chemistry*, 135, 140-148 (1998)

16. T. Vojnovich, R.J. Bratton, Impurity Effects on Sintering and Electrical Resistivity of Indium Oxide, *Amer. Ceram. Soc. Bull.*, **54**, N.2, 216-217 (1975)

17. U. Betz, M.K. Olsson, J. Marthy, et al, "Thin Films Engineering of Indium Tin Oxide: Large Area Flat Panel Displays Application", *Surface and Coating Technology*, **200**, N. 20-21, 5751-5759 (2006)

Author Index

Author Index